WEATHER
of BRITISH
COLUMBIA

Robin W. Pigott
Bill Hume

Lone Pine Publishing

The Publisher: Lone Pine Publishing
10145 – 81 Avenue
Edmonton, AB, Canada T6E 1W9

Website: www.lonepinepublishing.com

Library and Archives Canada Cataloguing in Publication

Pigott, Robin W., 1950–
    Weather of British Columbia / Robin W. Pigott, Bill Hume.

Includes index.
ISBN 978-1-55105-606-7

    1. British Columbia--Climate. I. Hume, Bill, 1949- II. Title.

QC985.5.B7P53 2009   551.69711   C2009-900316-3

Editorial Director: Nancy Foulds
Project Editor: Gary Whyte
Editorial: Wendy Pirk, Sheila Quinlan, Gary Whyte
Photo Coordination: Randy Kennedy, Gary Whyte
Production Manager: Gene Longson
Book Layout & Production: Rob Tao
Maps & Charts: Volker Bodegom, Rob Tao
Illustrations: Michael Cooke, Gerry Dotto, Megan Fischer, Trina Koscielnuk,
    Rob Tao
Cover Design: Gerry Dotto
Front cover photograph by: JupiterImages Corporation

Photo Credits: Every effort has been made to accurately credit copyright owners. Any errors or omissions should be directed to the publisher. The photographs in this book are reproduced with the kind permission of the copyright owners. All photo credits are located on p. 239.

We acknowledge the financial support of the Government of Canada through the Book Publishing Industry Development Program (BPIDP) for our publishing activities.

PC: 13

Table of Contents

Dedication

To my wife Laura. Her love, support and belief in me made this book possible.

Also to Bob Fortune, veteran Vancouver weather broadcaster, whose magical chalkboard drawings inspired my lifelong love of weather.

Acknowledgements

Many thanks to all the people who helped make this book come together. I would especially like to thank my co-author Bill Hume whose *Weather of Alberta* provided an excellent background text and model for this book. Bill's suggestions were invaluable and have made this a much better book with his input. The staff of Environment Canada in Kelowna and Vancouver, including Gabor Fricska, Allan and Lisa Coldwells, Chris Cowan, Pat Wong and in particular Jim Goosen supplied an extensive resource of photos, weather data and products. Jim's insightful review and many suggestions of material for the book are much appreciated.

Other agencies that contributed photos and information include the Canadian Avalanche Centre, BC Ministry of Forests and Range and Ministry of Environment.

Thanks to Mike Woodroff and John Mullock who spent many hours reviewing the text and Uwe Gramann, Shawn Pigott and Ed Tyson for their fantastic pictures. I am indebted to a former colleague of mine, Alan Whitman. His extensive database of tornadoes provides evidence of a more active storm occurrence in the province than would be expected.

The direction and guidance shown by Nancy Foulds, Wendy Pirk and the rest of the staff at Lone Pine Publishing is much appreciated. Finally, I would especially like to thank Gene Longson and David Cleary, former tennis partners of mine, for their suggestion that I be contacted to write this book.

Introduction

There's a saying in Alberta: "If you don't like the weather, just wait 15 minutes." Here in British Columbia, the phrase could be changed to: "If you don't like the weather, just move 15 kilometres." There are so many microclimates here in this province that, chances are, you can find the one that's right for you. Hourly or daily changes in the weather aren't usually that dramatic, but drive a few kilometres inland from the coast or climb a couple of thousand metres into the mountains and a whole different weather world awaits you. British Columbians like to brag that in spring you can ski in the mountains in the morning, then descend into the valleys and play tennis or garden in the afternoon.

On a larger scale, there is also tremendous variety in the weather, especially in winter when the contrast is sharpest. Imagine bitter cold and ice crystals suspended in the air in the Peace River country, and fresh and blowing snow on the Interior Plateau while Victorians are out counting flowers on a mild February afternoon. In summer, fog banks drift along the outer coast, the Fraser Valley cooks and thunderstorms rumble northward through the Interior.

British Columbia lies in the North Temperate Zone, which is the broad geographical area between the Tropic of Cancer and the Arctic Circle. The prevailing westerlies bring a constant stream of frontal systems and rain to coastal

Ecoregions of Canada

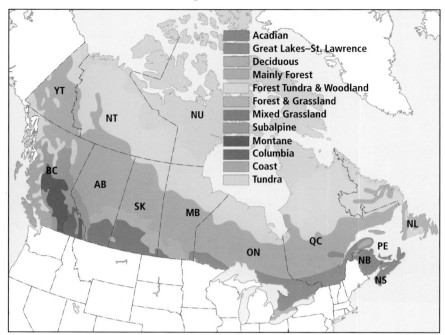

Legend:
- Acadian
- Great Lakes–St. Lawrence
- Deciduous
- Mainly Forest
- Forest Tundra & Woodland
- Forest & Grassland
- Mixed Grassland
- Subalpine
- Montane
- Columbia
- Coast
- Tundra

areas. As the air moves inland, each range of mountains extracts some moisture from the clouds, producing much drier conditions in the interior valleys and plains. The two general categories of climate are maritime and continental.

The varied topography and climate make BC the country's most biologically and ecologically diverse province. The biogeographic zones map on page 7 shows the distribution of forests and plants in the province. Coastal rain forests give way to drier interior forests. There are also, depending on elevation, differences in average annual temperature and amount of precipitation, as well as areas of alpine tundra, pockets of bunch grass and a vast boreal forest east of the Rockies.

The ocean has a moderating effect on the coastal climate, most significantly on the west coasts of Vancouver Island and the Queen Charlotte Islands. The result is a fairly narrow range of temperature differences throughout the year. Away from this influence, there are pronounced seasonal contrasts in temperature, especially over the Interior Plateau of the province. Cold, snowy winters and warm, fairly dry summers are typical of the continental climate.

Weather has played a huge role throughout the history of the province. First Nations peoples in the colder Interior had to be well prepared for arctic outbreaks in winter. In the milder coastal regions, it was a challenge to just keep dry, with incessant rains from autumn

Biogeographic Regions of British Columbia

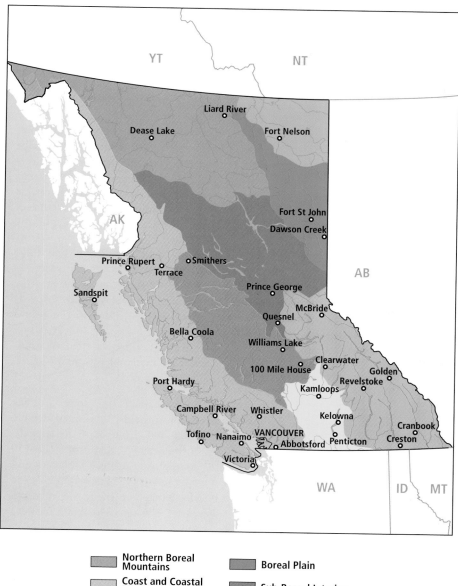

Northern Boreal Mountains

Coast and Coastal Mountains

Central Interior

Taiga Plain

Boreal Plain

Sub-Boreal Interior

Southern Interior Mountains

Southern Interior

through spring. European explorers, without any kind of weather forecast, were at the mercy of Pacific storms, and shipwrecks were common in the 19th and early 20th centuries. Travel inland through mountain passes was tough enough, but mud slides and avalanches from winter storms made it even more treacherous.

Today, satellite pictures, computer models, radar and a dense network of weather stations help forecasters predict the weather with a fairly high level of accuracy. However, even with this knowledge, the day-to-day changes in the weather, especially storms, still have a major impact on our lives. A recent example of this impact is the heavy snow event that occurred at the end of December 1996, which paralyzed the south coast for days. Interior travellers had to extend their Christmas holidays when all the mountain passes were closed because of numerous avalanches.

Speaking of extreme weather, many Canadian and even North American records for annual precipitation are set along the west-facing slopes of this province. Henderson Lake, almost midway

between Ucluelet and Port Alberni on Vancouver Island, tops all other wet spots on the continent—with a soggy 6650 millimetres of rain precipitation per year! Mt. Fidelity in Glacier National Park is the snowiest location in Canada, with 1471 centimetres each year.

Meteorology is the science that defines how the atmosphere behaves. This science is based on physics and mathematics, and the calculations used to describe the atmosphere are very complex. However, no matter how powerful a computer or model you have, it is still unable to give a completely accurate picture of what is going on in the weather at any given time. Therefore, there are errors from any starting position, and these errors magnify as the model generates maps and weather conditions into the future, which is why weather forecasts based on the model's predictions show decreasing accuracy with time. Read on to learn about the operation of the atmosphere and meteorology's many scientific complexities.

Types of cloud and cloud coverage across the province vary considerably each day and throughout the seasons. From the heavily laden rain clouds on the coast during winter to the dark and stormy thunderheads of summer in the Interior, there are many skyscapes to feast your eyes on. This book includes many examples and descriptions of clouds that give hints about the current state of the atmosphere and what to expect from the weather in the near future.

Winds can be destructive, but they can also power our generators and keep kites airborne. From hurricane-force winds on the coast, to severe thunderstorm gusts in the Interior, to Chinook winds descending from the Rockies, British Columbia definitely has its share of wind. Delve into these pages to read about the mechanics and types of wind in the province.

Because weather has such a huge impact our lives, people throughout time have kept their eyes on the skies. Weather observation is a science that has a long history; formal records in Canada date back to the mid-18th century. Some of the earliest records for British Columbia date back to 1879 in New Westminster, the early capital of the province. In this book, descriptions of the measuring systems and equipment used to observe weather in the province are provided.

Of the traditional basic elements, fire is by far the scariest. The summers of 1998 and, more significantly, 2003 were hot and dry over the Southern Interior. At the end of July and throughout August of both years, lightning strikes and human carelessness in parched forests started many fires that caused millions of dollars of damage to property. Fire weather meteorology attempts to describe and forecast the weather conditions that create the right conditions for fires. After fires are burning, detailed forecasts are written to ensure the safety of fire fighters. In these pages, learn about the Canadian Forest Fire Danger Rating System (CFFDRS) and the wildfire situation in BC.

Venturing beyond the day-to-day weather phenomena, the final chapters examine two broad atmospheric issues. In the first, we discuss air quality and learn how the atmosphere deals with air pollution. Second, we tackle climate change and the long-term future of the atmosphere, particulary how they affect British Columbia.

This book attempts to explain the operation of the atmosphere with narrative descriptions and diagrams, but it does not include any equations. It is hoped that the images included in these pages will help to illustrate meteorology's many scientific complexities. We discuss the atmosphere, its structure and how it works, with an emphasis on British Columbia and its varied climate. Those interested in a more rigorous treatment of the subject are better off referring to any number of scientific texts, some of which are included in a list of references at the end of the book. There are also many excellent weather websites. The Environment Canada site (www.msc.ec.gc.ca/weather) provides a lot of information on atmospheric science. Current British Columbia forecasts and computer model projections can be viewed at www.weatheroffice.gc.ca. By the way, if you google "weather" on the internet, more than half a billion pages show up!

Chapter 1: The Atmosphere

> **Some are weather-wise,**
> **some are otherwise.**
>
> —Benjamin Franklin

Compared to the dimension of the earth, the atmosphere is just a thin skin of gas. But it is complex; it contains a number of constituents, has a complicated structure and has many protective properties without which life as we know it could not survive. As we note in pictures taken from space, the blue atmosphere has considerable associated structure, which is visible because of the presence of clouds. When observed with time-lapse photography, movement and changes in the cloud patterns suggest that there is indeed some structure. There may be large areas with organized shape. White cloud masses swirl, expand and shrink, while cloud bands in the shape of lines and arcs move across the earth's surface.

Although the motion may appear chaotic, it actually closely follows paths and patterns that are described by rules governing the complex science of fluid dynamics. These are the same laws that describe the motion of fluids flowing in channels and pipes, or the motion of air around an aircraft wing. These motions all have a chaotic nature to them—they have some degree of associated turbulence. But the flow mostly behaves in an explainable manner, and its motion can be predicted.

The Composition of the Atmosphere

The composition of the atmosphere is 21 percent oxygen and 78 percent nitrogen. The remaining one percent includes water vapour, carbon dioxide, the inert gas argon and a wide variety of other gases, some naturally occurring and some human-made. Oxygen and nitrogen are well mixed throughout the atmosphere. Many of the other gases are less uniformly distributed, a result of their tendency to chemically react with other gases, to change phase or to remain in the region of the atmosphere near their emission source. Gravity keeps the gaseous envelope to within a depth of a few hundred kilometres of the earth's surface. The interaction between molecules pushes them apart, preventing the gas from collapsing into a lump of oxygen and nitrogen at the surface. This interaction is known as internal pressure. The molecules constantly collide with each other and with other objects. Near sea level, at temperatures near 20° C, the individual molecules move at about 500 metres per second. Molecules in the lower atmosphere are dense—near the earth's surface, 1 cubic centimetre contains about 25 sextillion molecules (or 25 followed by 21 zeroes!).

To picture an analogy for pressure, imagine yourself throwing balls at a bathroom scale. Each time a ball hits the scale, the scale registers a slight reading, depending on the speed and weight of the ball that was thrown. When many balls are thrown over a fixed period of time—one second, for example—the weight indicator on the scale does not have time to return to zero. A steady reading, or pressure of the colliding balls, is displayed. It is the same with the atmosphere. Over a 1-square-centimetre surface, there are hundreds of millions of

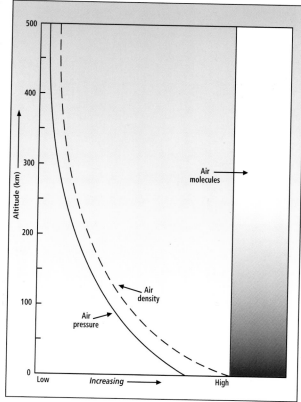

Variation of Atmospheric Pressure and Density with Height

Fig. 1-1

molecules, each with a tiny mass, striking per second. This molecular pressure pushing in all directions, including upward, counteracts the downward pull of gravity. The mass of all the molecules in a vertical column with a surface area of 2.5 by 2.5 centimetres extending from sea level to the top of the atmosphere is about 6.7 kilograms. This internal pressure, measured as a force per unit area, pushes back the earth's gravitational attraction to the molecules. In atmospheric science, the pressure force is called the millibar (mb).

The higher one goes in the atmosphere, the fewer molecules there are, and the weight of the molecules, as well as the corresponding internal pressure, is reduced. A mathematical equation defines how atmospheric pressure decreases with height; this equation shows (and measurement confirms) that at a height of 5 kilometres, atmospheric pressure is reduced to about 50 percent of its value at sea level, and at a height of 10 kilometres, to about 30 percent.

Seasons as a Result of Earth's Orbit

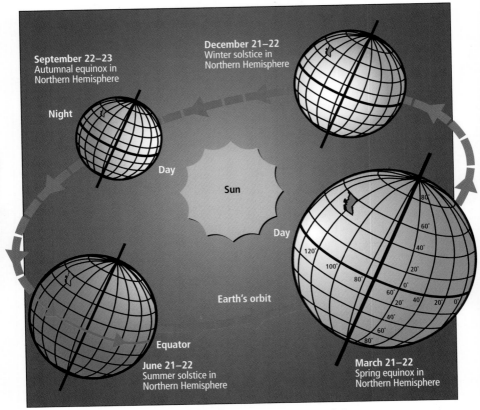

Fig. I-2 The tilt of the earth's axis of rotation gives us seasonal changes.

Solar Energy Reflected by Earth's Surfaces

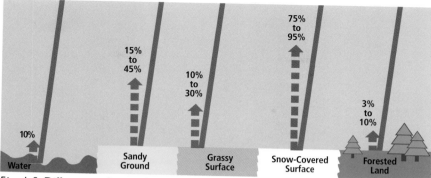

Fig. I-3 Different surfaces reflect varying amounts of the sun's radiation.

Solar Input and its Variation

The sun is the primary driver of weather on earth, but it is the tilt of the earth's axis of rotation that gives us the seasonal changes. This angle of inclination (23 degrees) remains constant as the earth proceeds on its annual trek around the sun. The orbit is not quite circular—it is elliptical. When the earth's axis is tilted away from the sun, and British Columbia is in the middle of winter, the earth is at its closest point to the sun.

The sun's radiation covers the full electromagnetic spectrum, from the very short x-rays and through the visible light rays to the much longer radio wave frequencies. However, the amount of radiation differs across the spectrum. About 40 percent of the sun's energy is emitted in the visible and near infrared frequencies. The amount of solar radiation that reaches the earth's surface is less than what is incident at the top of the atmosphere. Ozone molecules in the high atmosphere absorb most of the ultraviolet energy, while much of the incoming infrared energy is absorbed in the lower levels of the atmosphere.

Clouds and snow surfaces can reflect up to 95 percent of incident radiation, and water surfaces reflect about 10 percent. Sand, grass surfaces and forests reflect between 10 and 30 percent. These reflections depend on the angle of incidence of the radiation—the lower the angle of incidence, the greater the amount of reflection. Overall, 43 percent of the solar radiation entering the top of the atmosphere is reflected and scattered back to space. Some absorption by dust and gases, including water vapour, further depletes the incoming radiation, with the result that about 40 percent is absorbed by the earth's surface. All bodies that absorb energy are heated, and these warm bodies in turn emit radiation at the infrared frequency. This energy goes back out into space. There is a balancing act in constant play between the incoming and outgoing energy, and the average temperature of the earth is reasonably steady over time. Indications of an imbalance in this energy account give rise to the global warming concept, which is discussed later in this book.

Energy Transfers in the Earth-Atmosphere System

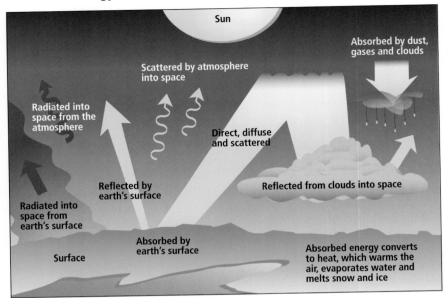

Fig. 1-4 Some solar energy is reflected and scattered back into space. The remainder is absorbed in the earth-atmosphere system, which radiates energy. Overall, it is almost in balance.

The British Columbia land surface plays a role in the global energy balance. Almost half of BC is covered by forest. There is also a large percentage of non-vegetated terrain in the numerous mountain ranges throughout the province. The remainder is a mix of ocean, lakes, urban areas and some grassland. Each surface has its rate of albedo (rate of reflectivity) that affects the absorption and reflection of solar radiation. For example, forests and water surfaces have a low albedo (they absorb almost all the incoming radiation); fresh snow has a high albedo, reflecting most of the energy back into space. Coastal areas generally have smaller changes in temperature, both between day and night and between seasons, mainly because of the moderating effect of the nearby oceans that absorb and slowly release heat. Also, with plenty of vegetation, outgoing radiation is limited at night, especially in winter when overnight temperatures stay warmer with the generally sparse extent of snow coverage. In the Interior, except near the larger lakes, there is more reflected energy at night. This effect is most pronounced in winter and early spring over snowy surfaces on calm, clear nights. Puntzi Mountain on the Chilcotin Plateau is a good example of this wide temperature variation. The weather station there sits at a fairly high elevation and lies in a bowl where cold air pools at night. One early April day back in the 1970s, the afternoon temperature rose to near 20° C with plenty of sunshine and warm, dry air descending from the Coast Mountains. At night, the snow cover acted as a good radiator of energy and the winds dropped to calm. Under clear skies, the temperature plummeted to a low of -20° C near dawn. The temperatures that day set a record for both the highest and the lowest temperature readings for a 24-hour period across Canada!

Vertical Variation of the Atmosphere

As already noted, atmospheric pressure decreases with altitude, but what about other measures? Temperature is the most significant parameter. The behaviour of temperature is used to define the layers of the atmosphere. Rates of temperature change with height are known as thermal lapse rates.

At British Columbia's latitude, temperatures in the lower atmosphere generally decrease from their highest values near the earth's surface to their lowest values at altitudes of about 10 to 12 kilometres. This layer of the atmosphere, characterized by decreasing temperature, or negative lapse rate, is called the troposphere, and it is here that all active weather occurs. The warmth at the bottom of the troposphere is a result of the warmth of the land

and water surface combined with the heat contained in water vapour. The temperature decrease stops abruptly or pauses at a level known as the tropopause, where temperatures are around -50° to -60° C.

Moving upward in the atmosphere, the temperature slowly increases through the layer known as the stratosphere, until it reaches its maximum at a height of about 50 kilometres. The stratosphere contains most of the atmosphere's ozone, and warming through this layer is mostly a result of the absorption of the sun's

Vertical Structure of the Atmosphere

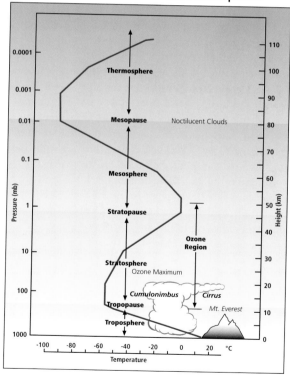

Fig. 1-5 Temperature changes with height and defines the different layers of the atmosphere. All the active weather is in the troposphere.

ultraviolet radiation. The maximum temperature is often in excess of 0° C at this height. This maximum temperature level is known as the stratopause. Thereafter, temperatures decrease again through what is known as the mesosphere, approaching a minimum value near -90° C at the mesopause, which lies at a height of approximately 80 to 90 kilometres and has an atmospheric pressure of about 0.01 mb. Above the mesopause, temperatures increase with height through a layer known as the thermosphere. In spite of low density within the thermosphere, the tenuous atmosphere can still affect drag on spacecraft at altitudes of 250 kilometres.

Moisture in the atmosphere plays a major role in the formation and development of weather. Virtually all the moisture is contained in the troposphere and is in its highest concentrations at the bottom of this atmospheric layer. The heating of this layer by the sun results in turbulent mixing, which distributes moisture and heat to higher levels. Moisture is also transported upward in the troposphere by the more gradual processes associated with what is known as frontal lift. Only rarely are the turbulent mixing or the frontal lift processes vigorous enough to drive moisture through the tropopause.

Convection starts in the lowest layer of the atmosphere, where warm air rises. As it moves higher, the air expands and cools. Nearby air sinks back toward the ground.

Atmospheric Circulation

Horizontal variation in atmospheric parameters has much more complexity than vertical variation. The earth's atmosphere can be thought of as a heat engine that is driven by the sun. In its simplest form, the basic function of the heat engine is to move heat from the areas where it is most intense—in the equatorial zone—toward the cooler regions near the north and south poles.

When scientists started to think about atmospheric circulation, they developed simple models to explain apparently organized phenomena. George Hadley developed a simple model in 1735 to explain the elements of the easterly trade winds that are observed in the tropics. Hadley noted that the equatorward motion of air in the easterly trade winds must be balanced in some fashion to prevent the accumulation of mass in the equatorial zone. He postulated that upward motion and then poleward motion of air would result. However, the simple model he proposed does not fully explain the circulation and associated weather patterns observed around the globe.

A model of atmospheric circulation developed by Carl-Gustaf Rossby in 1941 helps explain some of the complexity in the mechanism whereby heat is transported away from the equator toward the poles. This model, as depicted in Figure 1-6, has a considerable element of reality, especially in BC at so-called midlatitudes. In this model, the Hadley circulation shows flow toward the equator, ascent and return flow aloft, and then subsidence in the subtropics. This direct cell is driven by heat. The subtropics,

Hadley Circulation

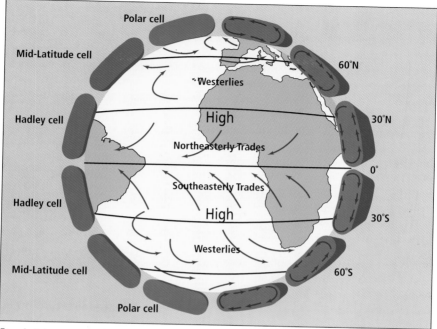

Fig. 1-6 A cross-section view of the atmosphere shows a three-cell circulation on each side of the equator. There is active weather in regions of general ascent.

where air is descending, are regions of generally high pressure (or anticyclonic circulation) and sunny skies, which are well known to the "snowbirds" who flee cold northern temperatures and head to those latitudes (from 20° to 30° N of the equator) for winter vacations. Another direct circulation cell is present at high latitudes (60° to 90° N). Here, cold air moves away from the pole toward the south. There is a general area of subsidence near the pole (surface anticyclone or high pressure), and a compensating northward flow aloft. Between the two direct cells is an indirect cell in which surface air is transported northward and air aloft southward. Warm air from the south and cold air from the north meet at mid-latitudes. This zone is where mid-latitude low pressure systems persist. This zone of air mass mixing, convergence and uplift associated with the mid-latitude low pressure systems provides all the key ingredients for production of the weather phenomena that we observe.

> **Blow winds and crack your cheeks! Rage, blow,**
>
> **You cataracts and hurricanes, spout**
>
> **Till you have drenched our steeples, drowned the cocks!**
>
> —Shakespeare, *King Lear*

17

The simple circulation processes described here do not take into account the rotation of the earth on its axis. This rotation has two important consequences. First, the zone of maximum heating moves from east to west as the earth rotates on its axis. Second, the earth's rotation also causes an apparent deflection of the airflow. On a rotating surface, an apparent force deflects any moving body. Gustave-Gasparde Coriolis first described this effect mathematically in 1835. North of the equator the deflection is to the right, and south of the equator the deflection is to the left. The magnitude of the apparent force is proportional to the speed at which the body moves.

Air parcels move because of the force of atmosphere pressure differences. Two bodies of air separated horizontally with different temperatures have different densities. Essentially, the colder, higher density air mass has a higher pressure than the adjacent warm, less dense air. Pressure differences between air masses cause air parcels between the masses to move toward lower pressure. This activity is known as the horizontal pressure gradient force. And the greater the pressure difference (the larger the pressure gradient), the greater the force and therefore the faster the air parcels move. At the same time, the Coriolis force causes a deflection of the air parcels. If we ignore other forces such as friction or the centrifugal acceleration in curved flow, the Coriolis force pretty well balances the pressure gradient force, and the air parcels move along lines of constant pressure.

Wind Deflection as a Result of Coriolis Effect

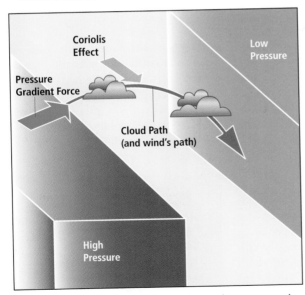

Fig. I-7 Air moving from high pressure to low pressure is deflected to the right in Northern Hemisphere.

Representation of Weather Systems

It turns out that the most convenient way to depict weather systems is by means of pressure patterns. Moving along a constant height surface, there is a reasonably large variation of pressure in the horizontal. For example, on a map drawn at sea level, pressure can vary by more than 100 mb. Atmospheric pressure is measured at many locations around the world. When pressures measured at the same time are plotted on a map, areas of high or low pressure can be discerned. An "L" marks the

Linkages Between Upper Air and Surface Patterns

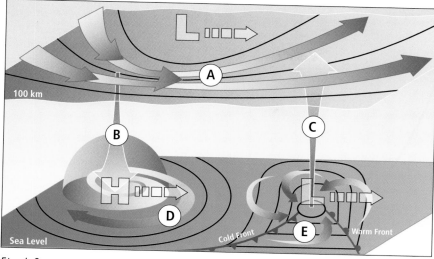

Fig. 1-8

A - wind flows around a moving low centre in the upper atmosphere
B - air subsides from an area of converging air aloft toward the ground
C - air ascends from a surface low toward an area of diverging air aloft
D - air spirals outward and clockwise around a surface high
E - air spirals inward and counterclockwise around a surface low

centre of an area with pressure lower than its surroundings, and an "H" marks the centre of an area of relatively high pressure. Lines joining locations that have the same pressure are called isobars. The greater the difference between the high and low pressure areas, the greater the number of isobars between those areas, and the greater the wind speed. So basically, a weather map is a depiction of pressure patterns with high and low pressure areas.

A number of other parameters are measured at the same time as pressure, and these parameters can also be plotted on the weather map. When the patterns

A synoptic chart is a map that displays the locations of highs, lows and fronts in a particular region.

of all these parameters are inspected, many relationships become apparent. Low pressure areas are generally associated with areas of cloud, and often with precipitation. High pressure areas are associated with relatively cloud-free areas. Relatively warm temperatures are usually near low pressure centres, and relatively cool temperatures are in areas of high pressure.

Maps of conditions at other levels in the atmosphere can also be drawn. When depicting weather patterns at upper levels, the relationship between pressure and height is used. Rather than depicting pressure patterns at a constant height, constant pressure levels are chosen and the height of that pressure level above sea level is depicted. Meteorologists examine pressure patterns at several pressure levels within the troposphere. Again, patterns in this height field can be analyzed. As one progresses upward in the atmosphere, the patterns become less "busy" than those at the surface. There are usually fewer centres of high or low height than centres of high or low pressure on a surface weather map covering the same area. As well, the patterns of height appear to have a rather uniform shape, often with broad areas that have an undulating or wave-like appearance. The relationships between these patterns, their shapes and the juxtaposition of patterns at different levels vertically then give the meteorologist a representation of the vertical structure of the atmosphere. Understanding the linkages between weather patterns through the depth of the troposphere is the essence of synoptic meteorology.

The study of synoptic meteorology has been underway for about 90 years. When surface observations were combined with observations aloft and the patterns were analyzed, it became possible to develop a three-dimensional model of how the atmosphere was structured. The knowledge that the atmosphere has structure and behaves in an understandable way led to a number of theories about how weather patterns moved and changed with time. The science is complex and can be represented by the mathematics of fluid dynamics, which we will not attempt to reiterate here—textbooks abound on the subject. But we can try to understand the elements of these atmospheric dynamics well enough to understand the formation, movement and prediction of weather systems over British Columbia.

Two key concepts play a crucial role in understanding the behaviour of weather in BC. The first concept is that of air masses and the so-called frontal regions that separate air masses. The second concept is the link between the motion of weather systems and the atmospheric flow patterns aloft.

Air Masses and Fronts

Air that remains over a land or ocean surface for some period of time takes on characteristics that depend on the underlying surface. In meteorology, the term air mass is used to describe a large pool of air that has a uniform character. If the land surface is intensely heated by the sun, the overriding air will be warm, and if the underlying surface is covered with snow or ice, the air will be cold. If the air is over an ocean, moisture will evaporate into the lower portion of the air mass. Air that remains over central regions of North America during summer is called

The jet stream was not discovered until World War II, when pilots noticed they lost ground speed when flying against it.

Fronts

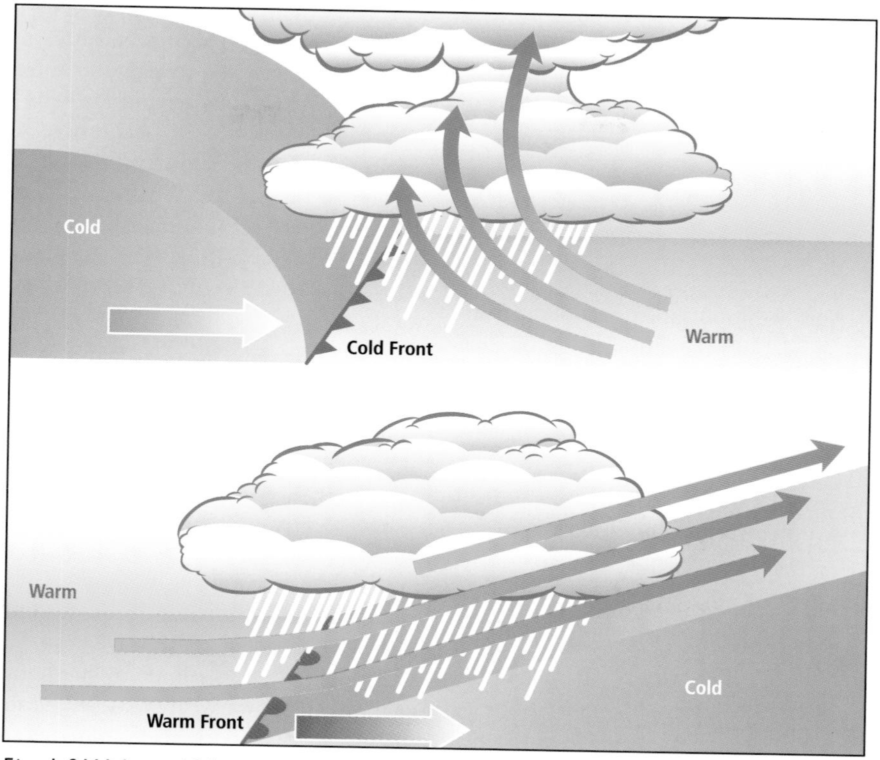

Fig. I-9 With a cold front, warm air is displaced by advancing cold, dense air and clouds are formed. With a warm front, warm air overruns the retreating cold air, forming clouds and precipitation.

a continental polar air mass, and we occasionally experience this air mass here. When the circulation is more active, air masses stream into BC from other regions. Air that originates from the northern Pacific Ocean, known as maritime arctic air, often crosses BC. Air from the arctic regions, which is dry and usually cold, is known as a continental arctic air mass. In winter, this air mass gives us the coldest (occasionally record-breaking cold) conditions. Maritime tropical air masses, originating over low

latitude regions of the Atlantic or Pacific Ocean, rarely come as far north as BC.

The boundaries between adjacent air masses are known as fronts. Between World War I (1914–18) and World War II (1939–45), Norwegian meteorologists established the frontal concept, as it is known, to give some structure to weather systems that influence mid and high latitudes. This concept is a key building block of modern meteorology. Weather conditions are usually described in relation to the presence, motion and modification of frontal zones. When air from a warm air mass moves against a cooler air mass, the boundary is known as a warm front. When the colder air mass prevails and pushes the warmer air, we have a cold front. As at other locations in the mid-latitudes, British Columbia's weather is frequently characterized by the approach and passage of warm and cold fronts.

Another important aspect of the frontal model of the atmosphere is that regions of low pressure frequently form along the frontal boundaries between air masses. The characteristics, behaviour and motion of these low pressure systems are the key to describing present and future weather conditions.

Lows bring strong winds, clouds and rain. Highs bring clear skies, perhaps with some high cloud, and can result in very cold weather in winter.

Motion of Weather Systems

The development and movement of frontal storms and weather systems is tied to weather patterns in the mid- to upper troposphere. The patterns differ between the northern and southern hemispheres, and there is little apparent interaction between patterns across the equator. As shown in Figure 1-10, the wind flow at these levels usually displays a wavy, undulating pattern. Meteorologists characterize these patterns as either long waves or short waves.

Long waves are known as planetary waves because they are discernable over the whole planet. The planetary wave, with its characteristic trough and ridge pattern, is continuous around the hemisphere. Over a period of days to weeks, there may be an apparently stable configuration in the planetary pattern with a distance of several thousand kilometres between successive troughs. These planetary waves usually move slowly eastward at a speed of about 10 to 15 kilometres per hour. Planetary waves will, on many occasions, appear to remain stationary, and portions of the pattern may even move to the west, a rare phenomenon known as retrogression.

Short waves, on the other hand, move relatively quickly. They appear as ripples in the planetary wave pattern and move along the long wave pattern from west to east. The speed of these waves can be fairly high, often 40 kilometres per hour or more. Short waves are closely tied to the so-called synoptic scale storms that are 300 to 500 kilometres across. In summary, short waves, with their attendant storm systems, move along, or are steered by, the planetary wave pattern.

The speed of the wind itself within the steering flow is usually much greater than the rate of progression of the short and planetary waves. As well, the amplitude of the short wave ripples usually decreases the higher we go in the atmosphere. The short waves are most apparent at the middle level of the troposphere; therefore, meteorologists pay a lot of attention to their shape, rate of motion and other characteristics. The long wave pattern is most evident at the top of the troposphere, and winds are generally

Wave Pattern in Wind Flow

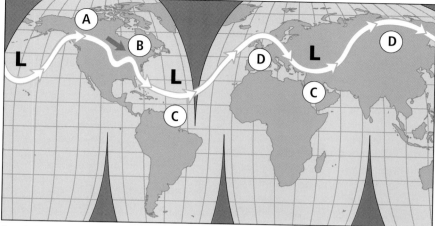

Fig. 1-10
A - jet stream
B - short wave moving through ridge

C - trough in planetary wave
D - ridge in planetary wave

strongest at this altitude. These winds are known as the jet stream. Meteorologists often designate the jet stream as the steering flow for short waves and storms.

The location of the ridges and troughs in the planetary wave pattern also play a role in determining average weather conditions. When the pattern is relatively stable, the associated weather conditions are also stable. Under a ridge, an area of high pressure dominates at the surface. High pressure areas are associated with subsiding, or a downdrift, of air. There is usually little cloud, and fair skies with sunny conditions are prevalent. On the other hand, a trough is associated with unsettled weather. When short waves round the bottom of the planetary trough, the associated weather systems are quite active. A considerable amount of vertical motion may be generated through the mid-troposphere, usually

producing cloud, stormy conditions and precipitation. The distance between the trough and ridge in the planetary wave pattern is often 2000 to 3000 kilometres. When the pattern is stable over Canada with the ridge positioned over the West, the trough is positioned over the East. As a result, warm dry conditions will prevail over British Columbia, while unsettled and perhaps cooler conditions will be prevalent in eastern Canada. Of course, the reverse situation also occurs.

Weather Influences in British Columbia

British Columbia is located within the mid-latitude temperate zone, where air masses and associated fronts are in a constant state of flux. With such a large change in latitude, there are often several air masses covering the province on any given day. It is not uncommon in winter to have bitterly cold temperatures from entrenched arctic air in the far north, snow falling in the Central and Southern Interior under a maritime arctic air mass and rain on the coast from much warmer maritime polar air. The upper level winds, and particularly the jet stream, play a part in the continual northward and southward migration of these air masses.

Any Grade 4 science student can tell you that the prevailing direction for mid and upper level winds in British Columbia is from the west. The moisture and relatively constant temperatures over the Pacific Ocean moderate the climate on the coast throughout the year. The north Pacific is a breeding ground for fronts and storms that continuously form, intensify and dissipate as they approach the coast. Just as no two snowflakes are exactly alike, each weather system has its unique characteristics; some bring heavy rain and strong winds, and others only produce scattered showers and light breezes. Between the disturbances, high pressure ridges bring sunshine and drying conditions that sometimes last for several days, especially in the summer months.

As the Pacific air moves inland, the temperature and moisture characteristics are modified by each rise and fall across the province. The windward sides of mountain ridges (i.e., the west-facing slopes) receive the highest amounts of rain and snow, while the east-facing down slopes into the valleys experience much drier conditions. With southwest winds aloft, the Olympic Mountains provide a rain-shadow effect, giving Victorians bragging rights as the driest spot on the south coast. The driest areas of the province are in the lee of the Coast Mountains. The village of Ashcroft annually records a meagre 150 millimetres of precipitation. The upslope effect results in wetter conditions over the eastern mountain regions; however, lower elevations, particularly the Rocky Mountain Trench and east of the Rockies in the BC Peace River area, are quite a bit drier with a westerly flow aloft.

Of course, the flow is not always from the west. Several times each winter, northerly winds bring cold, dry arctic air masses southward through the Interior and often all the way out to the coast. After a few days or sometimes weeks of unseasonably cold temperatures, the tap from the west gets turned on again, and rain or snow and warming temperatures return, first to the coast and then spreading inland.

Warm, humid airflows from the south also influence the weather here. In summer, after several days of hot weather, the upper ridge of high pressure starts to move eastward. This movement results in a southerly feed of subtropical moisture that often generates spectacular lightning displays and at times severe weather in the form of intense wind gusts and torrential downpours. The path of storms sometimes stretches all the way to the Central Interior.

The expression "go with the flow" is applicable in British Columbia. There are many other factors to consider, but a quick check of the direction of upper level winds gives you a first guess on what weather to expect each day.

Chapter 2: Clouds

For many people, clouds are the most interesting aspect of weather. Our perception of what each day will be like is based on our first glimpse of the sky and the presence or absence of clouds.

Clouds consist of tiny water droplets or ice crystals that occur as a result of the condensation of atmospheric water vapour. Condensation occurs with great difficulty in clean air; atmospheric water vapour must have a surface upon which it can condense. The formation of cloud particles therefore depends on the availability of tiny particles known as condensation nuclei. These condensation nuclei are almost always abundant in the atmosphere, and they arise from dust, pollution particulates, pollen and spores that are stirred into the atmosphere as a result of wind motion and turbulence. In most regions of the lower troposphere, the density is at least 10,000 per cubic centimetre. In industrial areas, concentrations can increase to more than one million nuclei per cubic centimetre. There is no shortage of condensation nuclei available for condensation!

There is an upper limit on how much water vapour can be present in the free atmosphere. This amount depends on the temperature and pressure of the air containing the moisture. Relative humidity is the primary way we express water content in the atmosphere. The relative humidity is a ratio of the amount of water vapour in the air compared to the maximum amount of water vapour that the air parcel can contain at the same pressure and temperature. This relative humidity is expressed as a percentage—air that is saturated with water vapour has a relative humidity of 100 percent. So, when clouds are formed, the relative humidity of the air containing the cloud is 100 percent. Another term that is often used to measure atmospheric moisture is the dew point temperature. When unsaturated air is cooled at a constant pressure and without the addition of moisture, the temperature at which the air becomes saturated and dew starts to form is called the dew point temperature. This temperature is often a value that is representative of the widespread characteristics of an air mass.

Cumulonimbus clouds can grow up to 18 kilometres high and can hold more than a half-million tonnes of water.

For cloud to form, there must be a mechanism for the air to increase its relative humidity to 100 percent. When air that contains water vapour cools, the vapour transforms into cloud droplets. Occasionally, fog forms when a moist air mass cools as it moves over a colder surface. Generally, however, air cools and expands as it ascends, moving into areas of lower pressure. The rate of cooling of the parcel in the lower half of the atmosphere is approximately 10° C per kilometre of ascent. Usually the water vapour remains within the parcel of air and cools at the same rate. When the water vapour cools to a temperature at which condensation starts occurring on condensation nuclei, the air parcel is described as saturated. Its relative humidity is 100 percent, and cloud forms. This altitude is then the base of the cloud.

If the parcel continues to rise, the rate of cooling changes. In fact, the condensing water vapour releases latent heat that was stored during the formation of water vapour by evaporation. This release of latent heat effectively cuts the rate of the parcel cooling by about half, which has important consequences for the development of convective clouds and thunderstorms.

Ascent of air can occur in a number of different ways.

Dynamic Lift

This lifting process results from the motion of the atmosphere at middle to upper levels in the troposphere. It is associated with the frontal lifting process, in that the ascent results from the relative motion of air masses. The ascent may be more widespread than that associated with the immediate frontal surfaces. Generally, broad scale ascent is associated with the counter-clockwise motion around a low pressure system. Descent of air, or subsidence, is associated with clockwise flow around a high pressure system.

Orographic Lift

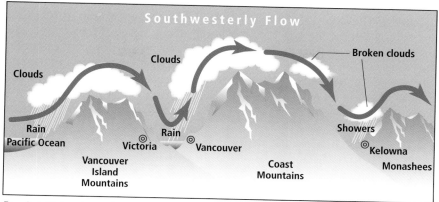

Fig. 2-1 Moist air moving eastward forms clouds on west-facing slopes. Clouds dissipate to the lee of higher terrain..

Orographic Lift

Air can be forced to ascend as it moves along sloping terrain in a process known as orographic lift. Because of the presence of mountains and sloping plains in British Columbia, orographic lift is a very relevant process here. It is predominant in the lower atmosphere, though if the mountains are high enough, as with the Rocky Mountains, the lifting process can extend well into the upper troposphere.

Frontal Lift

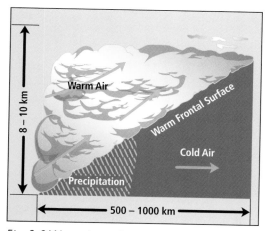

Fig. 2-2 Warm air moving upward along a frontal surface

Frontal Lift

A warmer air mass ascends over a colder, denser air mass. The interface between the two air masses is known as the frontal surface. One process that often governs ascent of air is associated with atmospheric frontal surfaces. This process is in many ways similar to orographic lift, except the lifting agent is the frontal surface that surrounds the colder air rather than the earth's surface. Frontal surfaces can extend through the whole troposphere, so frontal lift can occur through a considerable depth.

Convection

One mechanism that produces ascent is convection. This process is common during daylight. As the sun heats the land surface, air adjacent to the land warms. The amount of warming depends on the type of land surface and its aspect ratio to the incident rays of the sun. Some air parcels are a little warmer than adjacent parcels, and they rise. As they rise, they cool. If these parcels are warmer than the adjacent air at that level, they continue to rise. This process continues until the rising air parcels reach a level at which the surrounding air mass is warmer. At that level, the upward motion of the parcel ceases.

Turbulent Lift

This mechanism is caused by the friction between a moving air mass and the underlying surface. Turbulent lift is somewhat similar to convection except that air parcels get their upward motion from undulations in the underlying surface. The height to which the turbulence extends depends on the speed of airflow (i.e., wind speed), the roughness of the underlying surface and the stability of the air mass itself. Turbulent lift generally does not extend far above the earth's surface, but it can be horizontally widespread.

Convective Bubbles

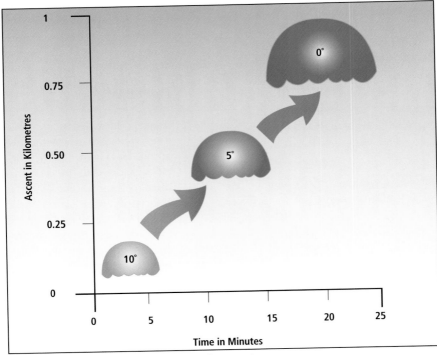

Fig. 2-3 Parcels of air rising as a result of convection

Cloud Classification

An international classification system for clouds has been developed for use by ground observers. Clouds are generally classified into four families based on how high the cloud base is located above ground level. Low clouds have bases in the layer that reaches up to approximately 2 kilometres above ground level, though cloud based at ground level is generally called fog. Middle clouds are based between about 2 and 6 kilometres above ground level and are known as alto form (Latin for "middle"), while high clouds are based above 6 kilometres. Clouds that have a considerable vertical dimension and that are based in either the low or middle range are called cumulus (Latin for "heap").

Each cloud family is subdivided based on appearance. In general, the visual appearance of clouds is closely associated with the process that gives rise to the lifting process. Clouds that appear as a layer and are spread out horizontally are called stratus clouds (Latin for "layer"). Most high clouds are wispy in appearance, largely because they consist predominantly of ice crystals that are formed at the low temperatures associated with the upper troposphere. These clouds are known as cirrus (Latin for "curl of hair"). Latin terms are used to further define appearance (fractus, castellanus, pileus, lenticular, mammatus) or other cloud characteristics, such as the presence of rain. These terms are often used in combination. The nimbus subdivision is used for cloud types from which precipitation is falling.

What follows is a detailed description of the cloud types seen in British Columbia.

Stratus

As observed from the ground, these clouds appear as a uniform layer usually resulting in an overcast condition—that is, the sky is uniformly covered from horizon to horizon. These clouds are usually made up of liquid water droplets, but if the temperature at cloud level is cold enough, stratus clouds can contain a large number of ice crystals. Usually these clouds do not have any permanent markings on their lower surface, and individual cloud elements are indistinct. Precipitation from these clouds is rare and is at most a drizzle or light snow.

Stratus Fractus

These clouds have a ragged appearance and are based at the same level in the atmosphere as a stratus deck. Stratus fractus clouds result when the conditions that formed a stratus cloud layer are no longer active and the stratus deck breaks up.

Stratocumulus

Stratocumulus clouds consist of a series of patches of rounded clouds that have relatively little vertical development. These clouds are often associated with stratus clouds and are either a precursor to the formation of stratus or result from the transition of a stratus deck into cumulus cloud. The deck of stratocumulus clouds is not continuous; the edges are ill defined, and the cloud has a patchy or rolling appearance. Blue sky or higher cloud decks frequently appear through breaks in the layer of stratocumulus. Any precipitation associated with these clouds is at most a drizzle or light snow.

Fog

Fog can be considered a cloud that is in contact with the ground. It can form when a moist air mass flows over a colder surface, or in calm wind conditions when moist air in contact with the ground cools overnight to the point that its relative humidity rises to 100 percent. When fog breaks up, usually in the presence of a slight breeze, a layer of stratus clouds may form above the ground. As ground heating

increases in conjunction with an increase in surface wind, the base of the stratus deck may rise. Fog may also dissipate because of the action of sunlight, which heats the top of the fog layer. Some of the solar energy also penetrates the fog deck, warming the underlying ground surface and raising the air temperature beyond the point of saturation.

Nimbostratus

As the nimbo preface implies, this cloud is often associated with continuous wide-spread rain or snow. This cloud is generally quite thick, extending from 1 to 2 kilometres above the ground at its base to tops often over 5 kilometres. The cloud base is usually dark grey to black, and cloud elements are indistinct.

34

Helmholtz Waves

Helmholtz waves resemble breaking waves on water. These clouds are formed along a horizontal interface between warm air that is flowing over colder air. They are rarely seen because there is not usually enough moisture at that interface for clouds to form.

TYPE OF CONDITIONS CLOUDS USUALLY BRING:

cirrus — *usually means fine weather*
cirrocumulus — *the mackerel sky is sometimes an indication*
   of unsettled weather
cirrostratus — *approaching rain or snow*

Cloud Streets

Low clouds frequently align along the direction of low level wind, resulting in what is known as a cloud street. These clouds are generally stratocumulus or cumulus. A steady horizontal wind initiates a rolling motion in the flow, with gentle ascent combined with gentle subsidence occurring along parallel lines. A line of cloud forms in the areas of ascent.

altostratus — *rain or snow likely if cloud thickens*
nimbostratus — *rain or snow*
stratus — *maybe light rain or drizzle*
cumulus — *sunny days*
cumulonimbus — *thunderstorms, showers and sometimes hail*

Altostratus

Altostratus has a grey to steely blue colour and generally covers the whole sky. Thin layers of altostratus frequently resemble ground glass. This cloud has no characteristic thickness and can be up to 6 kilometres deep, which is thick enough to completely obscure the sun. The base of the cloud appears flat when observed from ground level, but when flying though an altostratus layer, the lower boundary is virtually indistinct. Precipitation rarely falls from these clouds.

Altocumulus

This cloud type is similar to stratocumulus. The altocumulus cloud deck is not usually widespread. The cloud elements are arranged in groups, and they frequently form into lines, usually following the direction of the wind at their formation level.

Altocumulus Castellanus

These clouds usually develop within a line of altocumulus. They are an indication that the middle atmosphere is unstable. When they form in the mornings during summer, they occasionally grow to be quite large vertically, and precipitation can fall from them. Occasionally they even develop into weak thunderstorms.

Lenticular and Wave Clouds

These clouds are often observed in the vicinity of hills and mountain ranges, which makes BC a good place to see them. When air that is thermally stable vertically is forced to flow over obstructing terrain, it frequently forms a standing wave. Standing waves along the foothills of mountains are well known to glider pilots, who use them to provide the lift necessary to soar to heights above 10 kilometres. The moisture flowing through the wave pattern condenses in the updraft portion of the wave. Where the air subsides, cloud dissipates, giving the cloud the form of a curved lens. Because the waveform can often remain in position for a long time, the cloud is usually stationary in the sky for as long as the upper level winds and moisture conditions are favourable.

Lee Wave Wind Pattern with Lenticular and Rotor Clouds

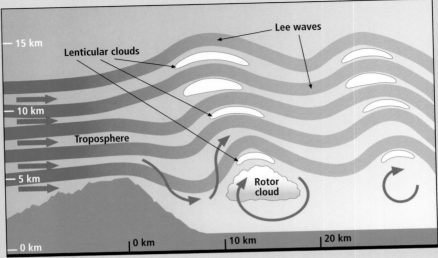

Fig. 2-4 A cross-section of wind/flow associated with standing wave pattern

Standing Waves and Chinook Arch

To the downwind (lee) side of mountains, standing waves can have a lengthy axis parallel to the mountain range. When looking toward the mountains from ground level, the clear sky framed by cloud to the east gives the curved appearance of an arch. This arch is most often observed in northeastern BC in the lee of the Rockies but at times can be seen to the east of the Coast Mountains in the Chilcotin.

> A cloud does not know why it moves in just such a direction and at such a speed...It feels an impulsion...this is the place to go now. But the sky knows the reasons and the patterns behind all clouds, and you will know, too, when you lift yourself high enough to see beyond horizons.
>
> —Richard Bach

Rotor Clouds

These clouds occasionally accompany wave clouds in the lee of ranges of hills. Rotor clouds are quite rare in BC but can occasionally be observed in the lee of the Rockies. Precipitation does not usually accompany rotor clouds, and they are most notable for extreme air turbulence. They are therefore of great concern to low-level aircraft operation.

Standing lenticular cloud

THE COLOUR OF CLOUDS

You can often tell what is going on inside a cloud by what colour it is. Storm clouds are dark because they have a high water droplet content and the droplets are tightly packed together, so light has trouble passing through them. Thunderstorm clouds take on a green tinge because light is scattered by the ice in the cloud. If you see a green-tinged cloud, you can be pretty sure that heavy rain and hail are on the way. A yellow cloud gets its colour from the presence of smoke, usually from forest fires. Yellow clouds are rare but can sometimes be seen during late spring or early autumn, when forest fires are most common.

Cirrus

Cirrus clouds consist entirely of ice crystals. These clouds can take on a variety of forms but generally appear as fibrous wisps showing white against the blue sky. Isolated tufts occasionally have feather-like plumes that stretch out and turn upward. They are often called mares' tails.

Cirrostratus

Also made up of ice crystals, cirrostratus clouds appear as a thin, whitish veil that can extend from horizon to horizon, giving the sky a milky appearance. Occasionally there is a fibrous appearance to the cirrostratus layer, which implies that there are streaks of thicker strands of cirrus cloud embedded within the layer.

Halos are sometimes seen around the sun or the moon as it shines through the cirrostratus layer. This phenomenon is often thought to be an omen of bad weather, which has some truth. A cirrostratus layer is formed from the overriding of moist air along an elevated frontal surface, which indicates that a different air mass is approaching, heralding a change in the weather.

45

Cirrocumulus
These clouds have a patchy or, occasionally, wave-like appearance. They create a slight shadow, if any.

Noctilucent Cloud

Noctilucent clouds occur in the mesopause at an altitude of about 80 kilometres. These clouds are made up of ice crystals, and the condensation nuclei are thought to be meteoric dust particles. The moisture content at this level is patchy, so the clouds are rarely observed. Because these ice crystal clouds are thin, they are only observed while the sun is still below the horizon. Noctilucent clouds have no effect on weather and are more of an observational curiosity. We are at the prime latitude in British Columbia (from 50° to 60° N) to see them, but because layers of lower cloud block the sky, noctilucent clouds are not often seen here.

Contrails

Aircraft exhaust contains large amounts of water vapour and particulate matter. The exhaust gas cools quickly and, with the cold temperatures at flight level, usually results in the formation of ice crystal clouds that appear to stream out behind the aircraft. The turbulent motion of the atmosphere causes these contrails to spread laterally, and the turbulence created by the aircraft means contrail dissipation usually occurs within a few minutes. However, if the atmosphere at the contrail level has a high relative humidity, the contrails may persist for a while longer. The horizontal motion of the atmosphere at aircraft flight level further complicates the phenomenon. While successive aircraft fly along the same path relative to the ground, the air at flight level is almost always moving, which gives the impression that successive contrails are side by side. The presence of successive and perhaps long-lived contrails may result in the formation of a thin cirrus shield. This shield will grow throughout the day, and in areas that are frequently crossed by over-flying aircraft, a fairly extensive deck of cirrostratus can result.

Clouds with Vertical Development

The temperature of the ambient air that a parcel of rising air encounters governs how high a cloud can grow. If the temperature of the air within the parcel is warmer than the surrounding ambient air, the parcel continues to rise. When the air contained within a rising parcel becomes saturated, the latent heat energy is released. The cooling rate of the saturated air is reduced to about one-half of the cooling rate of the unsaturated parcel. Because the surrounding ambient atmosphere generally cools at a rate between the moist and dry rates, the rising saturated air parcel remains warmer than the surrounding ambient air. In this situation, the atmosphere is said to be convectively unstable, and clouds experience rapid vertical growth.

Cumulus

These clouds generally have a flat base, and they are as high as they are wide. The edges of the clouds are usually quite distinct, giving the clouds the puffy, white appearance of popcorn. These fair weather cumulus clouds, often known as cumulus humilis, generally go through a standard life cycle. In late morning, the intensity of solar heating reaches a point where parcels of air near the surface rise, forming puffy cumulus clouds (we say the "cumulus starts popping"). As the day progresses, the appearance of the individual clouds slowly changes, and they may seem to be relatively motionless in the sky. As sun intensity decreases late in the day, the clouds generally dissipate. Precipitation rarely falls out of these cumulus clouds.

Towering cumulus clouds

Towering Cumulus

The cumulus cloud base may be 1 to 2 kilometres above the ground, and the tops of towering cumulus may extend to heights of 4 to 5 kilometres. These clouds contain rather vigorous updrafts and often produce precipitation in the form of rain showers or snow flurries. Towering cumulus clouds may become more organized and can grow to a significant extent horizontally as well as vertically. These larger groups of towering cumulus are known as cumulus congestus.

Occasionally, the tops of towering cumulus will give an upward push to a layer of moist air aloft. This moist air condenses, forming a veil-like layer of ice crystal cloud (essentially cirrus). These veiled cumulus congestus clouds are known as cumulus pileus.

Towering cumulus clouds (above), cumulus congestus clouds (below)

Cumulonimbus

The cumulonimbus cloud represents the ultimate growth phase of cumulus cloud into a vigorous, precipitating cloud mass. Cumulonimbus clouds can stretch horizontally for kilometres and extend from their base through the depth of the troposphere, to heights of 10 to 15 kilometres. Lower portions of these clouds contain liquid moisture (cloud and rain droplets), while upper portions consist of ice crystals. The upward growth of these clouds is constrained by the thermal stability of the tropopause. The tropopause acts as a lid, and the tops of cumulonimbus then spread laterally, forming a fibrous cirrus veil. Strong winds aloft generally move the veil downwind, resulting in the classic anvil-like appearance of the cumulonimbus cloud. The anvil top can extend many kilometres downwind of the main cloud base. This cloud almost always generates showery precipitation, and more vigorous storms produce hail within the cloud. This hail frequently reaches the ground.

Fully Developed Cumulonimbus

Turbulence

Anvil Top

Storm movement

Cold
Downdraft

Warm
Updraft

Roll Cloud

Wind Shear
Turbulence

Wind Shear
Turbulence

Rain

Gust Front

Fig. 2-5 Air that spirals into the base of a cumulonimbus cloud can form a wall cloud.
Air pushes out the top of the cloud to form a cirrus anvil.

Chapter 3: Precipitation

Water vapour is present even in the driest of air masses, but it is the conversion of vapour into liquid or solid forms that creates much of the weather we experience. A simple analogy for how water vapour results in precipitation can be made with a sugar-saturated cup of hot coffee. When the solution cools, the liquid cannot continue to retain all of the sugar in its dissolved state, and some precipitates out as solid crystals.

Water vapour behaves like any other gas in the atmosphere, but atmospheric water has the ability to convert between solid, liquid and gaseous phases. The atmosphere cannot hold an unlimited amount of water vapour, and the amount of vapour that can be held entirely depends on the temperature of the mixture. In the atmosphere, if the air at a given temperature is saturated with water vapour, cooling results in condensation into water droplets and sometimes sublimation directly into ice crystals.

Air does not contain a large amount of water vapour. One cubic metre of air at sea level, with a temperature of 35° C, weighs about 1 kilogram. If that cube is saturated with water vapour, the water has a weight of about 37 grams. We can measure atmospheric water vapour content in a number of different ways. Vapour pressure is the actual pressure exerted by water vapour, and it is always a small fraction of the air pressure. At a given temperature, there is a maximum vapour pressure that can be reached before a phase change occurs. This process is known as the saturation vapour pressure.

Condensation in the atmosphere does not happen readily because the air molecules are always in motion. To stick together and form a liquid, the molecules require the presence of condensation nuclei. These nuclei are small, ranging from 0.1 to 10 microns (thousandths of a millimetre) in diameter. The lower atmosphere contains between ten thousand and one million condensation nuclei per cubic centimetre. In the upper atmosphere, there are hundreds

per cubic centimetre. For ice crystals to form, freezing nuclei are required. However, freezing nuclei are far less numerous than condensation nuclei. There may be only one or two per cubic centimetre.

Condensation forms cloud droplets that are typically 10 to 20 microns in diameter. When raindrop sizes are measured, they typically have a diameter of about 2000 microns. So, a typical raindrop is made up of more than one million cloud droplets. A major challenge for atmospheric scientists was to determine what process occurs in the atmosphere to form raindrops from cloud droplets.

Official Definitions of Precipitation

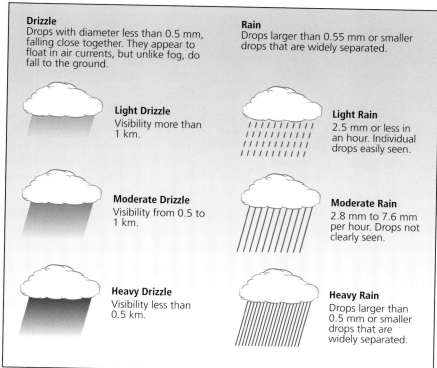

Drizzle
Drops with diameter less than 0.5 mm, falling close together. They appear to float in air currents, but unlike fog, do fall to the ground.

Rain
Drops larger than 0.55 mm or smaller drops that are widely separated.

Light Drizzle
Visibility more than 1 km.

Light Rain
2.5 mm or less in an hour. Individual drops easily seen.

Moderate Drizzle
Visibility from 0.5 to 1 km.

Moderate Rain
2.8 mm to 7.6 mm per hour. Drops not clearly seen.

Heavy Drizzle
Visibility less than 0.5 km.

Heavy Rain
Drops larger than 0.5 mm or smaller drops that are widely separated.

Fig. 3-1

The Raindrop Formation Process

Precipitation forms in the presence of water vapour if cooling of air also happens. For most cloud formation, this process usually occurs when a parcel of air is lifted. This lifting can happen when an air parcel is heated near ground level and becomes less dense than neighbouring air parcels. Another lifting process occurs when air moves over elevated terrain. Also, if an air mass moves horizontally against a cooler air mass, the air masses do not readily mix, and the warmer air rides over the colder air along the boundary between the air masses (the front). This lifting process is known as frontal lift. The rate of cooling also depends on whether or not the air parcel is saturated with moisture. If the air parcel is not saturated, it cools at a rate of about 10° C per kilometre. A water-saturated parcel cools at a rate of 5° to 7° C per kilometre.

Two processes are thought to lead to the formation of raindrops. In the 1930s, scientists Alfred Wegener, Tor Bergeron and Walter Findeisen developed a theory about a process that depends on the presence of ice crystals. Ice crystals and water droplets form at temperatures below freezing, but when a mixture of water droplets and ice crystals is present, the ice grows more rapidly than the water droplets. In fact, water droplets do not spontaneously freeze until temperatures are below -30° C.

So, at cold temperatures, ice crystals spontaneously form and grow at the expense of water droplets. Gravity causes these ice crystals to fall. The cloud droplets also fall, but at a much slower speed. The fall speed of droplets and ice crystals depends on what is called their terminal velocity. As they fall, the crystals continue to grow in the water-saturated air, and they collide with cloud droplets that have slower terminal velocity. These droplets immediately freeze when they strike the ice crystals, and splinters of ice may be ejected, which also grow into droplets. The ice crystals fall and, assuming that their fall speed is greater than the upward speed of the surrounding air,

Growth of Ice Crystals

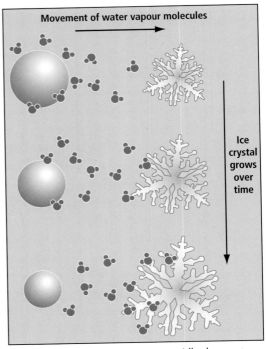

Movement of water vapour molecules

Ice crystal grows over time

Fig. 3-2 Ice crystals grow more rapidly than water droplets when water vapour is abundant.

Raindrop Growth by Accretion

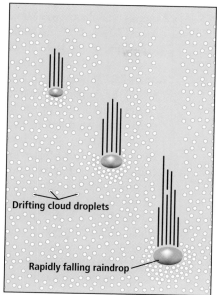

Drifting cloud droplets

Rapidly falling raindrop

Fig. 3-3

they move toward the ground. When they pass into air that is above freezing, they melt and form raindrops. This process is the "ice crystal theory," also known as the "Bergeron process," and seems to explain the formation of snow and most rain.

The ice crystal theory, however, does not explain the formation of rain in clouds that do not contain temperatures below freezing. In the tropics, and even occasionally at mid-latitude locations such as BC, precipitation falls from warm clouds. A second theory known as the "collision theory" explains this rain formation process. In this theory, the terminal velocity of cloud droplets of different sizes has to be considered. Larger cloud drops fall faster and overtake smaller droplets, and the raindrops grow by accretion. With the warm cloud accretion process, the rate of raindrop formation and the size of the resulting drops are directly proportional to the depth of the cloud. In British Columbia, the precipitation formed in warm clouds usually consists of small drops that are known as drizzle.

Of course, the two precipitation-forming processes are usually underway at the same time in the atmosphere. In clouds that have significant depth, such as nimbostratus or cumulonimbus, both processes are happening within the cloud. If there are two distinct layers of cloud, snow and ice crystals that form in a layer with temperatures below freezing, such as cirrostratus, may fall into a lower layer of cloud that is, in part, above freezing. This snow can stimulate production of raindrops in the lower cloud, which normally may not be thick enough to form raindrops.

A lot of rain starts off as snow, high in the atmosphere. It melts as it falls through warmer air near the ground, becoming rain.

Types of Hydrometeors

Rain is any liquid precipitation that reaches the ground. Raindrops can range in diameter from 0.5 to 7 millimetres. They are spherical in shape if they are less than about 2 millimetres in diameter, but larger raindrops falling through the atmosphere are flattened along the bottom because of the upward force of air. This force causes drops larger than 7 millimetres across to break up.

Snow Shapes: Basic Structure

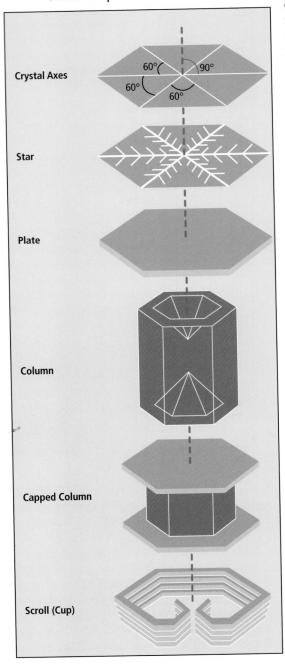

Drizzle consists of small drops that are less 0.5 millimetres in diameter. These drops fall out of relatively thin cloud decks and can be associated with fog. Drizzle drops have slow fall rates, and even the slightest breeze causes drizzle to appear to swirl. If temperatures are just below freezing at ground level, drizzle drops freeze on contact. The accumulation of ice from freezing drizzle can be significant and can result in treacherous travel conditions on our highways and hazardous conditions for aircraft. Freezing drizzle conditions occur mostly in northeastern BC, especially during autumn and winter.

Snow crystals come in a variety of shapes and sizes. Snow in the atmosphere forms shapes that depend on the nature of the water molecule. The water molecule, which consists of one oxygen and two hydrogen atoms, has a polarity based on the positive and negative charge distribution. When water molecules link together in an ice lattice, opposite electrical polarities of the atoms attract, forming a solid ice crystal that has a hexagonal symmetry. The hexagonal building

Fig. 3-4 The different shapes depend on the crystal structure of ice and the relative growth rates on each crystal axis.

block grows in one of many characteristic fundamental shapes, which makes it possible to categorize snow crystals into several different classes. Figure 3-4 shows this crystal structure and the shapes of the various snow crystals and flakes than can result.

It is probably true that no two snowflakes are alike. As snow falls toward the ground, it can take on a variety of shapes. Perfectly symmetrical snow crystals are rare. Usually the air is at least somewhat turbulent, and the branches of stellar flakes can break off. Snow crystals may break as they strike each other, or they may stick together. Cloud droplets also strike the snowflakes and may freeze instantly, forming what is known as a rime coating. This process causes the original crystal or flake to take on a lumpy, irregular shape.

As one would expect, average annual snowfall in British Columbia is highest over the interior mountain ranges and lowest along the coast and through the inland valleys. In some warmer winters, the south coast of BC

Fig. 3-5 Temperature is the primary factor that controls shape, but available water vapour also has an influence.

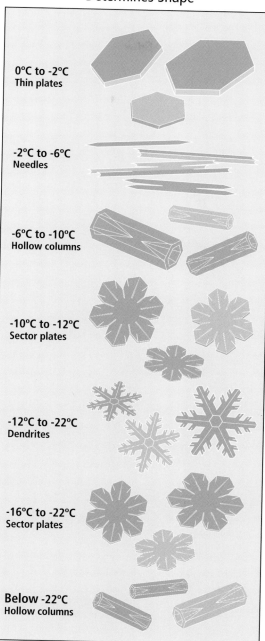

Snow Shapes: Temperature Determines Shape

0°C to -2°C
Thin plates

-2°C to -6°C
Needles

-6°C to -10°C
Hollow columns

-10°C to -12°C
Sector plates

-12°C to -22°C
Dendrites

-16°C to -22°C
Sector plates

Below -22°C
Hollow columns

How Graupel and Hail Form

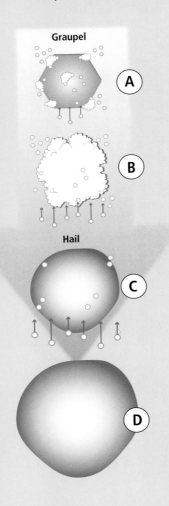

Fig. 3-6
A - supercooled cloud drops freeze to the crystal
B - cloud and raindrops freeze to form graupel
C - hail forms as more liquid raindrops freeze
D - hailstone falls out of the cloud when the updraft can no longer support its weight

records little or no snow at all. The high mountains in eastern BC receive the most annual snowfall. Snow falls on more than 140 days each year, the highest number of snow days in the country. The honour for most snow in any town nationwide goes to Stewart, at the head of Portland Inlet on the north coast, where around 570 centimetres accumulate each year.

When ice crystals are subjected to a lengthy riming process as they fall through deep cloud, they form **graupel** or **snow pellets**. These heavily rimed ice particles are sometimes called soft hail, and they are usually no bigger than 5 millimetres. They may take on a conical shape. Snow pellets can be seen at ground level when temperatures are at or slightly below freezing. When they hit a hard surface, they may bounce but often shatter. Formation of graupel is important in the cloud electrification process. Graupel formation in deep convective clouds is the main mechanism for the formation of charged regions that lead to the generation of lightning.

Very large graupel samples compared to a dime, a nickel and a quarter.

Formation of Ice Pellets and Freezing Rain

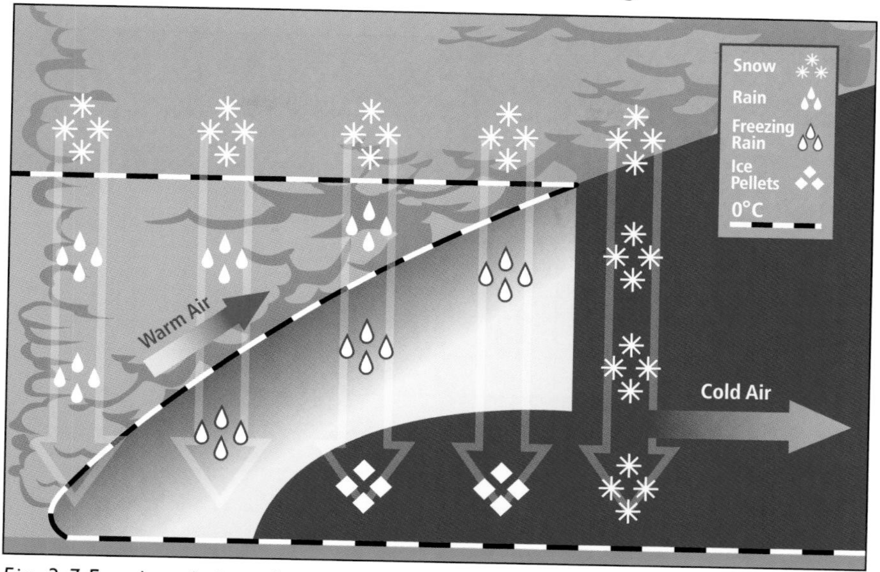

Fig. 3-7 Freezing rain, ice pellets or snow depend on the thickness of the sub-zero air mass.

Snow grains are small, opaque particles of ice that consist of bundles of rime or ice crystals held together by frozen cloud droplets. They fall at a light, steady rate from layers of relatively thin cloud and are considered to be the solid equivalent of drizzle. They usually have diameters of less than 1 millimetre.

Ice pellet is the name commonly given to precipitation consisting of refrozen raindrops or refrozen, partially melted snowflakes. The particles of ice are usually transparent, or nearly so. Ice pellets as observed at ground level are formed when raindrops refreeze after falling through a layer of cold air near the ground surface. They are 1 to 5 millimetres in diameter.

When there is a significant layer of warm air aloft and a thin layer of cold, sub-zero air at the ground surface, we get **freezing rain**. In this case, the surface-based cold air is not deep enough to freeze the falling raindrops, and they freeze when they strike objects at ground level that have a temperature below freezing. Such situations are relatively rare in BC, but when they happen, they can disrupt transportation. Airports, in particular, are keenly attuned to the possibility of freezing rain because a coating on runways can shut down airports for lengthy periods if a prolonged cold spell follows a freezing rain event.

> **Getting an inch of snow is like winning ten cents in the lottery.**
>
> —Bill Watterson, *Calvin and Hobbes*

Top BC Freezing Rain Sites

City	Canadian Ranking	Number of Days
Prince George	75	5.26
Fort St.John	76	4.60
Williams Lake	81	2.90
Chilliwack	82	2.87
Cranbrook	86	2.33
Abbotsford	87	2.30
Kelowna	89	2.06
Kamloops	90	1.34
Vancouver	91	1.15
Penticton	92	0.97
Courtenay	93	0.73
Vernon	94	0.68
Prince Rupert	95	0.62
Victoria	96	0.54
Duncan	97	0.49
Nanaimo	98	0.44
Port Alberni	99	0.24

Fig. 3-8

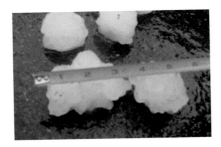

Hail, though uncommon in BC, can at times be very destructive to property and crops. In their simplest configuration, hailstones are more or less spherical and are made up of concentric layers of clear and opaque ice. They form within actively growing convective shower clouds. Usually the electrical charge separation process is also underway in such clouds, so most hail occurs in association with lightning.

In active and developing convective clouds, strong upward and downward air movement can exist side by side. An individual hailstone begins its life as graupel that is supported by the strong updraft in the convective cloud. In strong updrafts, the graupel particle may rise a significant distance—perhaps several kilometres above its level of origin. As it moves upward, it passes through regions that have high moisture content and a dense concentration of supercooled cloud droplets. With the high rate of collisions between the droplets and graupel, freezing is slowed for part of the accreted liquid, and the graupel becomes impregnated with water that freezes relatively slowly. The resulting pellet is more or less transparent. The growth of a hailstone is illustrated in Figure 3-6.

The hailstone is buoyed by the strong updraft and can grow to a significant size as the water-saturated air flows past. In upper regions of the cloud, the updraft speed may reduce or the hailstone may move into an area with reduced uplift. When the stone becomes too heavy to be supported by the updraft, it begins a downward trajectory toward the earth's surface, all the while sweeping through supercooled droplets and continuing to grow. Often, the hailstone falls back into the stronger updraft and again moves upward. Figure 3-9 illustrates how hail growth occurs in a mature thunderstorm cell. Updraft regions can be 1 kilometre or more across, and speeds in the most intense regions have been measured reaching 30 metres per second or more.

Structure of a Hailstorm

Fig. 3-9
A - hailstones grow in weaker updraft and fall out as medium-sized hail
B - large hail rises higher in cloud in strong updraft region

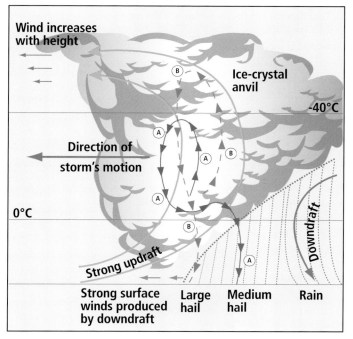

Hailstone Cross Section

Different layers in a hailstone reveal different temperature and water vapour conditions at each growth stage.

The largest hailstones grow in these environments. Large hailstones often have a lumpy structure, possibly because the hailstones are spinning and tumbling in the updraft, and water streams off the stone, forming icicle-like projections. Many hailstones have an onion-like structure with many layers that formed as the stone moved between different updraft regions with different temperatures and liquid water content. This structure is revealed when cross-sections of hailstones are examined, as shown in the photo above. Extremely large hailstones often consist of smaller individual stones that have collided and become frozen together within the storm cloud.

Hail in British Columbia

Hailstorms can occur anywhere across the province. They are more frequent on the coast, but the stones are generally small and may be reported as hail though they are more likely ice or snow pellets. Hailstorms are most often observed in the winter months, typically in the unstable air behind active frontal systems. Because of their small size, damage from the stones is always minor, if any. Although hail is not as common in the Interior, the stones are larger and are usually associated with severe summer thunderstorms. The highest hail frequency is in the Peace River country, with only a few storms each year producing hail on the Interior Plateau. Although hail may last for only a few minutes at a specific location, the storms move along tracks that may be hundreds of kilometres long.

Virga

This phenomenon deserves mention because it is often seen in dry environments, such as the Southern Interior in the warmest months of the year. It is defined as streaks of rain or ice particles falling from the base of a cloud but evaporating before reaching the earth's surface. This phenomenon is frequently observed over inland areas of British Columbia in summer because precipitating clouds are often based well above ground level, and the evaporation process is enhanced in the dry air at lower levels over the province. The presence of virga can be associated with dry downbursts, which themselves are a product of the cooling of air as it evaporates. In such cases, the presence of virga can indicate a hazard to aviation. Incidences of virga have also been misinterpreted as funnel clouds associated with thunderstorms.

Virga can sometimes have the appearance of a funnel cloud.

Chapter 4:
Atmospheric Electricity

Lightning

The atmosphere has both an electrical structure and a physical structure. In the lower portion of the atmosphere, lightning plays a key role as an electrostatic generator that helps recharge two concentric conductors—the surface of the earth and the electrosphere, which consist of highly conductive layers at altitudes above 50 to 60 kilometres. In the upper atmosphere, above 80 kilometres, the electrical phenomenon of most significance is the aurora borealis, which results when particles emitted by the sun interact with the atmospheric gases at those levels. It is thought that there is no interaction between the electrical processes in the lower and upper atmospheres.

The earth's surface has a negative electrical charge, and the lower atmosphere has a positive charge. In fair-weather conditions, this positive charge flows steadily to the earth's surface. In fact, if nothing else happened, the earth's charge would be neutralized in approximately 10 minutes, depending on the presence of conducting polluting gases. The thunderstorm acts as a generator that maintains the positive electrical charge in the atmosphere by means of the lightning discharge.

Thunderstorm Generator Charges the Earth-Atmosphere Battery

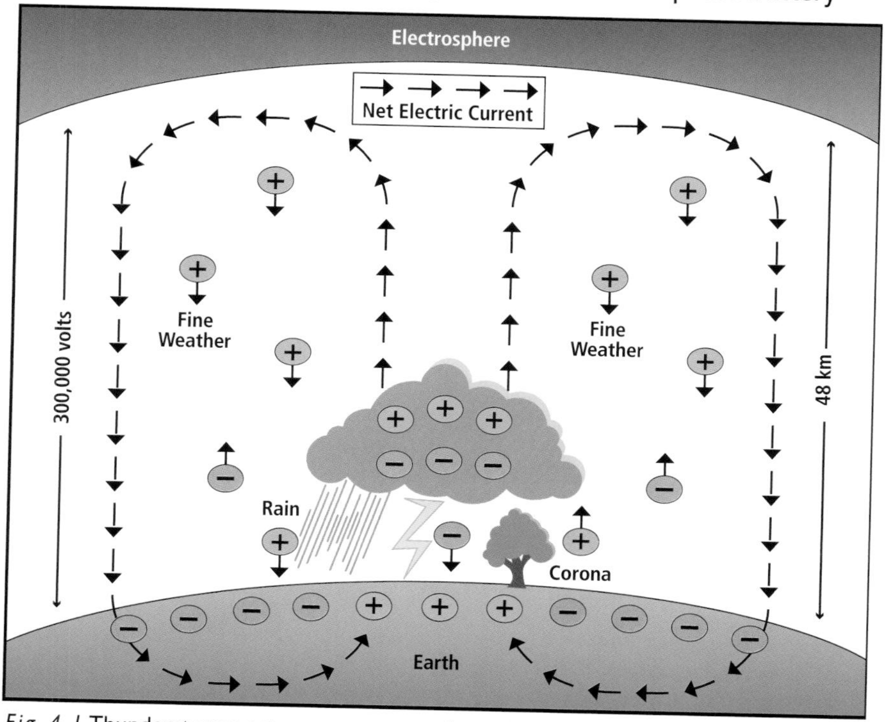

Fig. 4-1 Thunderstorms act as generators to keep the earth charged negatively and the atmosphere charged positively.

Around the world there are approximately 2000 thunderstorms in progress at any given time, and with the rate of cloud-to-ground lightning flashes in an average thunderstorm, there may be 30 to 100 flashes to ground every second. In British Columbia, the thunderstorm season is generally from May to September, though thunderstorms sometimes occur outside of this period. During summer, there is usually at least one thunderstorm in progress every day.

The Empire State Building in New York City has been hit by lightning up to 500 times in one year; it was once hit 12 times within 20 minutes. Toronto's CN Tower is hit 40 to 50 times per year. So much for "lightning never strikes the same place twice."

Convective Mechanism of Cloud Electrification

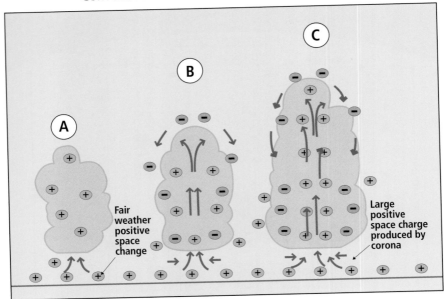

Fig. 4-2
A - positive charge swept into a developing cumulus
B - negative charge attracted toward edge of cloud
C - downdrafts on edge of cloud transfer negative charge to cloud base

The Cloud Electrification Process

Two mechanisms are thought to be responsible for electrifying clouds, turning them into thunderstorms—the convection mechanism and the graupel-ice mechanism. Although the latter process is the most widely favoured, the two processes may work in tandem, and there may even be an electrification process that is still unrecognized.

The Convective Process

Convective clouds go through a growth and intensification process that was described earlier. Convective clouds consist of updrafts of air parcels that are buoyant relative to the adjacent air mass.

These rising parcels contain moisture that condenses, forming the visible cumulus cloud. In the convection theory, shown in Figure 4-2, external sources provide the electrical charge. Updrafts at

With lightning, the "leader" stroke is the one that reaches from the cloud to the ground, setting up a path for the "return" stroke, which travels from the ground up to the cloud. It is the return stroke that we actually see.

the cloud base carry positive, fair-weather charges toward the top of a growing cumulus cloud, and the charge is more concentrated than that in the surrounding air. Negative charges produced by the constant bombardment of cosmic rays are attracted to the positive cloud and, in turn, attach to the cloud particles at the outer portions of the cloud. These negatively charged cloud parcels move downward in the convection circulation at the edge of and outside the cloud, compensating for the convective updrafts inside the cloud. This negative charge moves toward the cloud base where it can discharge to earth, producing a positive corona, which produces a positive charge under the cloud and a positive feedback process. This process does not explain all the complex charge distributions in fully developed thunderstorms.

The Graupel-Ice Process

The graupel-ice process cannot function unless a precipitation process is present in a growing cloud. The precipitation elements are graupel particles. As gravity draws graupel particles toward the earth, they collide with cloud droplets and ice crystals that are at the same levels within the cloud. When collisions occur, the graupel particles, as well as nearby ice crystals, acquire either a positive or a negative charge. The sign of the charge depends on the temperature of the air, the water content of the cloud, ice crystal and water droplet sizes, the relative speed of particles in the collisions and the chemical pollutants in the water.

The Graupel-Ice Charge Separation Process

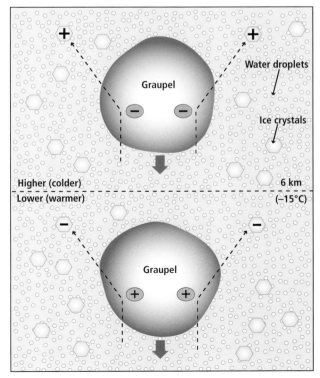

Fig. 4-3

Charge Distribution in a Thunderstorm Cloud

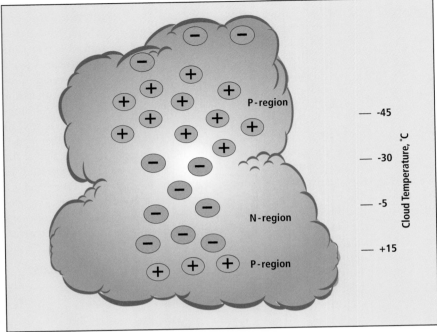

Fig. 4-4 The likely charge distribution in a fully charged thunderstorm

The most important factor is the temperature at which the collisions occur. In the upper part of the cloud where the temperatures are colder, the falling graupel acquires a negative charge and the crystals acquire a positive one. At warmer temperatures, the graupel is positively charged and the adjacent ice crystals are negatively charged. The critical temperature is between -10° and -20° C.

This charge-separation process then results in what is known as the tripole distribution observed in thunderstorms (see Figure 4-4). In this tripole, a positive charge is located at the top of the cloud (known as the P-region), a large region of negative charge is in the lower half of the cloud (known as the N-region) and a smaller region of positive charge (called the p-region) is at the cloud base.

Lightning can heat the air it passes through up to 30,000° C. The surface of the sun has a temperature of about 6000° C.

Coronal Discharges and Lightning Forms

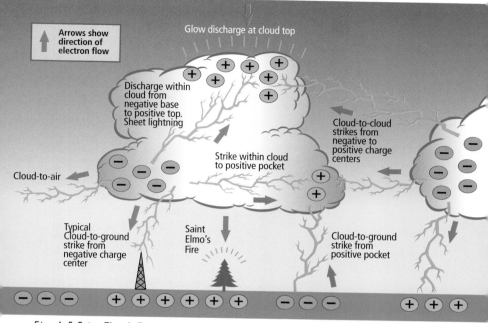

Fig. 4-5 Saint Elmo's Fire is a coronal discharge that occurs when a thunderstorm is present. Many different types of lightning occur in thunderstorms.

Generation of a Lightning Stroke

There are two primary types of lightning stroke: cloud-to-ground strokes (from the charged portion of the cloud to the earth) and intracloud strokes (between charged portions of a cloud). Occasionally, strokes pass between charged portions of two clouds (called cloud-to-cloud lightning), and even more rare is a strike from cloud to air. During thunderstorms in North America, there are about five times as many intracloud discharges as cloud-to-ground discharges. Most of the pictures of lightning strikes are of the cloud-to-ground variety because the intracloud flashes are usually obscured by the cloud itself.

Although the charged cloud typically has a tripole configuration, the N-region is of greatest importance in cloud-to-ground discharges. As the cloud moves over the land surface, this negative charge region induces the build up of a positive charge at the earth's surface. This positive charge area stays under the cloud as the cloud moves along.

The lightning stroke mechanism is complex, but most scientists believe the process starts when the negative and highly mobile electrons in the N-region discharge to the small p-region immediately below. The electrons overrun the p-region and continue their trip toward the ground in a rapid series of bursts

A cloud-to-ground lightning stroke

The upward movement of the luminous region moves at a speed of about 100,000 kilometres per second, or about one-third of the speed of light, and takes about 100 millionths of a second. This process is known as a single stroke discharge. The human eye is not capable of resolving the rapid speed of the luminous channel, and the flash appears to be one continuous stroke. Typically, however, several strokes occur within the same channel, progressively discharging higher and higher portions of the N-region. These consecutive strokes are typically separated by gaps of only 40 to 50 thousandths of a second. This rapid sequence of flashes gives lightning its flickering appearance.

The lightning discharge process is, of course, what is responsible for thunder. The rapid flow of electrons in the lightning channel causes a rapid heating of the channel to about 30,000° C. This heating causes an explosive expansion of the channel. This expansion compresses the surrounding air, producing a shockwave that propagates rapidly in all directions. After only a short distance, the shockwave converts to a sound wave that propagates at the speed of sound, which near sea level is about 330 metres per second.

along a channel known as the stepped leader. The lower end of the stepped leader moves toward the ground at about 120 kilometres per second. As the tip of the stepped leader approaches the ground, a strong positive charge is induced, especially on objects projecting above the ground surface. Upward-moving positive discharges, called streamers, are initiated from these points, and soon the downward stepped leader and the upward positive streamer connect. Electrons at the bottom of the stepped leader rapidly discharge to the ground, and the lower portion of the channel becomes luminous. As electrons are pulled from higher and higher up the channel, the region of the strongest surge of downward rushing electrons also moves upward, as does the associated luminous region.

Anatomy of a Lightning Stroke

Fig. 4-6
A - electrons accumulate in the base of the cloud
B - electrons move downward along stepped leader
C - streamer moves upward from an object on the earth's surface
D - upward moving positive charge creates a luminous discharge channel

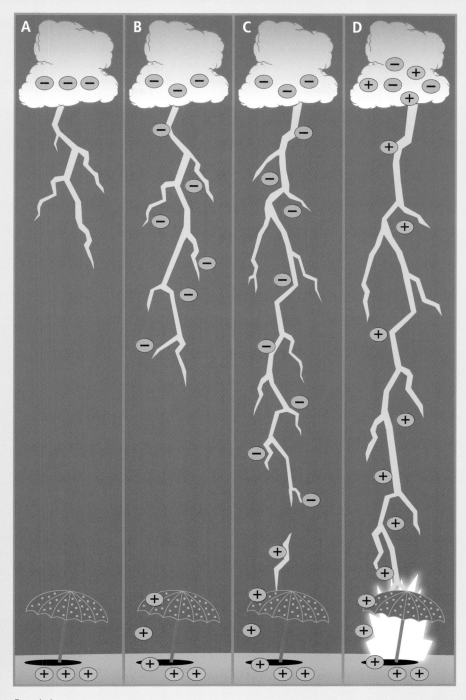

Fig. 4-6

For an observer, the light from a flash appears instantaneous. However, the sound moves much slower. For a cloud-to-ground discharge, we hear the sound from the closest portion of the channel first and from the most distant portion last. Therefore, depending on the orientation of the lightning discharge channel, the thunder duration may be spread out over several seconds.

The sound of thunder is a result of lightning rapidly heating the air it passes through, causing the air to expand explosively.

The pitch of the thunder depends on the intensity of the explosive heating and the altitude at which the thunder shockwave is generated. To add complexity, the atmosphere itself filters out higher pitched sound at a rate that primarily depends on the distance the sound has to travel through the air. Nearby thunder has a relatively high pitch, while more distant thunder has had a lot of the higher pitches suppressed and is a low rumble. In a nearby lightning discharge, we occasionally hear what sounds like hissing or cloth tearing. That sound is believed to be a combination of the sounds made by the downward-moving stepped leader and the upward-moving streamers.

The lightning process described has been concerned with discharges from the N-region in thunderstorms, but occasionally when the P-region is closest to ground level, it initiates a discharge. This discharge can happen in a number of ways: if the cloud is strongly sheared so that the lower N-region is dispersed; if the cloud is relatively shallow and only develops a small N-region; or if the cloud is dissipating and the residual upper portion of the cloud remains intact. Thunderstorms forming in winter also have a higher frequency of positive lightning discharges because, in cold weather, the charge formation process favours establishment of a more extensive P-region. Lastly, a positive discharge can occur from a thunderstorm passing near a tall tower or elevated terrain, such as a nearby mountain peak.

Positive discharges typically consist of a single stroke, rather than the multiple strokes that are normally experienced. The positive strokes also carry a large amount of current to the ground. The magnitude of the total charge exchanges can be up to 10 times larger than those in a typical negative discharge. With these large current flows, positive strikes may also have a greater tendency to cause combustion and may be a primary cause of forest fires, especially because they often take place well away from a precipitating region of the storm cloud.

Although most lightning is of the cloud-to-ground or intracloud variety, upward discharge has also been observed from the top of the thunderstorm. There have been reports, mainly by high-flying U-2 aircraft, of upward propagating lightning strokes that have the same appearance as commonly observed lightning. These discharges appear to be a few kilometres in length.

Multiple strokes in a time-lapse photograph of a thunderstorm

> **Thunder is good, thunder is impressive; but it is lightning that does the work.**
>
> —Mark Twain

Other, more diffuse discharges have also been observed. Known as red sprites and blue jets, they extend 10 to 30 kilometres above large thunderstorms. They appear to occur in association with significant cloud-to-ground strikes. Although rare, their presence had been proposed as early as the 1920s. Sprites and jets are thought to form when a pulse of energy moves upward at the same time as a downward lightning discharge. This energy pulse, in turn, causes electrons to break free of gas molecules in the region above the thunderstorm.

A cascade of energized electrons moves upward and causes the oxygen and nitrogen molecules to emit light. A ring of green light may also appear to flash at altitudes of 80 to 100 kilometres above ground level in association with the red sprite.

These phenomena are difficult to observe because clouds are usually blocking the view of a storm's top, and the actual discharge is faint. If they are seen at all here in BC, sprites and jets might be observed well past sundown with active thunderstorms that are 100 to 200 kilometres away, most likely in flatter areas such as the Interior Plateau or east of the Rockies in the Peace River country.

Ball lightning in a 19th-century woodcut

Ball Lightning

Small, luminous spheres sometimes observed during thunderstorms are called ball lightning. These balls are reported to be about the size of an orange or grapefruit, though some have been reported to be as large as 30 to 40 centimetres in diameter. The balls are usually located within a few metres of the ground. They move erratically, occasionally appearing to bounce along the surface. Lightning balls mostly appear yellow to red and seem to be rather innocuous. When they disappear after a short life span of seconds to one minute, they usually decay silently, though there have been reports of a small explosive sound. At present, there is no reliable theory for the formation and behaviour of ball lightning.

Areas of Lightning in British Columbia

Lightning detection networks, established to assist in forest fire detection and management programs, have made it possible to map the location and frequency of cloud-to-ground and intra-cloud lightning strikes. In BC, lightning can be observed somewhere in the province on most summer days. With its generally cooler temperatures, the coast does not receive strikes very frequently. However, in the Interior, the strong daytime heating combined with unstable air produces hundreds of strikes on many days. Several times each year, warm moist air moves up from the south, and all it needs is a trigger in the form of a cold front to set off many thousands of strikes. If the forests are dry enough, many new fires are started.

Effects of Lightning Strikes

Forest Fires

Given the global rate of cloud-to-ground strikes, forests worldwide might be expected to receive about 50,000 strikes per day. Not all strikes initiate forest fires, but the BC Ministry of Forests estimates that lightning is responsible for starting close to 60 percent of the wildfires in the province. During the 11-year period ending in 2007, there was an average of 1060 wildfires each year initiated by lightning in BC. Studies of the characteristics of a typical cloud-to-ground stroke suggest that peak electrical currents are in the range of 10,000 to 20,000 amperes, with occasional peaks over 100,000 amperes. This current flows for a short period—

usually less than one-thousandth of a second. Often there is a continuing current of 100 amperes or so that lasts for another one- or two-tenths of a second, and this type of stroke, with a short continuing current, is believed to be responsible for starting combustion. Lightning appears to follow a path through a thin layer of living cells between the inner bark and the wood layer of a tree. With the passage of the large currents through the tree, there is explosive heating—a strip of bark is generally blown off, and the tree itself may be split or splintered. Depending on the dryness of the tree bark and ground cover, a fire may be ignited. In many cases, the fire does not start immediately, and combustible material below ground level may smoulder for a few days before surface combustion begins.

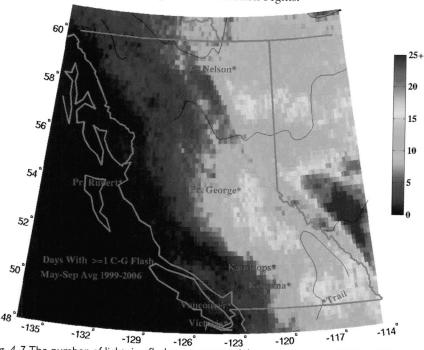

Fig. 4-7 The number of lightning flashes per square kilometre per year, 1999 to 2006

Fulgurites can form when lightning strikes sandy surfaces (below).

Fulgurites

When lightning hits a land surface that is made up of sand and certain kinds of rock, the heat can be high enough to melt the material along the lightning channel. This material solidifies, creating what is known as a fulgurite. Fulgurites take the form of long, hollow tubes that range from 1 to 5 centimetres in diameter. They have been traced in surface sand layers to depths exceeding 15 metres. Ancient fulgurites that date back more than 250 million years have been uncovered.

A lighting bolt can contain up to 100 million volts of electricity.

Lightning Protection

Given the destructive nature of the cloud-to-ground lightning stroke, protecting property from lightning-initiated fire has a long history. Benjamin Franklin undertook experiments confirming that lightning had an electrical nature, and he was probably the first to suggest, in 1750, that lightning rods should be placed on structures to guide lightning to the ground through a conducting wire. Today, lightning codes have been established that outline the characteristics of an adequate protection system for residences. As well, high voltage electrical transmission lines have lightning protection systems built in. Combustible fluids must also be protected to ensure that lightning or any related electrical arcing does not contact any air-vapour mixtures.

Passive Electrical Phenomena

A variety of other processes occur in the atmosphere, the results of which can be observed in the sky in BC. These processes occur at various layers of the atmosphere and have a number of causes. They, for the most part, result in the production of a weak, light glow and can be classified as having an optio-electrical basis.

Coronal Discharge

Coronal discharges occur any time there is a strong electrical field present in the atmosphere. Such strong electrical fields occur underneath active convective storms and cause ionization of the air molecules. When the gas is ionized, electrons in the molecule are elevated to a higher energy level. When the electrons return to their stable state, they emit light—a process known as fluorescence. This process is similar to the mechanism that causes neon lights to glow. In the atmosphere, the coronal discharge appears as a luminous bright blue or violet glow that can be accompanied by a hissing or buzzing sound.

Electric fields are more concentrated in areas of high curvature, such as at the end of a lightning rod, ship mast, church spire, chimney or other pointed object. Aircraft flying through active electrical storms often develop coronal discharge streamers from antennas and propellers, and even from the entire fuselage and wing structure. The discharge can also appear on leaves and grass, from the tops of trees or thunderstorm clouds and even at the tips of cattle horns. Coronal discharge, also known as St. Elmo's fire, has been associated with ships at sea during stormy weather (Elmo is the patron saint of seafarers).

Lightning Safety Tips

In the open
- stay at least 30 metres away from metal fences
- remove shoes that have metal cleats
- do not use metal objects such as bicycles, golf clubs or fishing poles
- do not shelter under trees or canopies or in small sheds, picnic shelters and the like
- avoid open fields and high ground; if you are in an open field, crouch down and cover your ears
- seek shelter in low-lying areas, such as valleys or depressions, that are not prone to flash floods

Indoors
- keep windows and doors shut, and do not approach them
- do not have a bath or shower or use tap water—electricity can be carried through the pipes
- unplug all electrical appliances
- do not use a phone that is connected to a land line

In a vehicle
- you are safe in an all-metal vehicle as long as you do not touch anything on the interior that is metal
- do not park near trees or under power lines
- if a power line should fall on or near your vehicle, do not get out of the vehicle
- convertibles are NOT safe—it is a vehicle's outer body that makes it safe, not the rubber tires

Aurora Borealis:
The Northern Lights

All the electrical effects discussed to this point are associated with phenomena in the lower atmosphere. The aurora borealis occur at a much higher altitude; the dazzling displays of light and colour have no apparent influence on the weather we experience at the earth's surface, and they are sometimes put in a category known as space weather.

The earth has a strong magnetic field that is produced by the movement of the earth's molten core, and this magnetic field streams out into space. The sun is continually emitting a stream of positively and negatively charged particles known as plasma gas, which moves outward in what is often called a solar wind. Although the earth's magnetic field is greatly distorted by the pressure of the solar wind, the magnetic field serves the important function of keeping highly energized solar particles from reaching the earth's surface. The upper reaches of the earth's magnetic field also contain electrically charged plasma gas. The theory of the formation of the aurora borealis holds that the flow of the sun's plasma past the earth's plasma causes electrons to become highly energized. These electrons strike the rarified atmospheric gases more than 80 kilometres above the earth's surface. These collisions impart energy to, or excite, the gas molecules. When the excited gas molecules return to their stable state, they emit light energy, which is the source of the aurora borealis. The colours we see are characteristic of the types of gases that are present in the atmosphere and depend on the energy of the particles that are stimulating the gas molecules. Atomic oxygen

emits green and dark red light, and nitrogen emits blue and purple. The colours can be varied, and they shift depending on the motion of the magnetic field and the incident speed of the solar ions.

The shape of the auroral pattern is directly related to the shape of the magnetic field at that level. The aurora borealis generally have a fairly sharp lower cutoff at their base height of 80 to 100 kilometres. The top of the pattern is generally indistinct, but it can extend several hundreds of kilometres above the base. The aurora borealis are quite spectacular and in British Columbia are most often seen during winter. Their frequency is greatest at about 60° N, so the number of events and intensity of auroral displays improve the farther north you are located.

81

Another impressive lightning stroke

It has been suggested that the aurora borealis emit sound; however, the region of excitation is located at least 80 kilometres above the surface, and the tone of the discharge has a fairly high frequency, based on what is observed with fluorescence discharge sounds in the lower atmosphere. The atmosphere preferentially absorbs high frequency sound waves, so it is doubtful that the sound would reach the earth's surface.

> **The weather is like the Government, always in the wrong.**
>
> —Jerome K. Jerome

Air Glow

Air glow is a weak, blue-coloured emission by the atmosphere that forms as a result of several processes in the high atmosphere. During daylight hours, solar rays can cause molecules in the upper atmosphere to become ionized, and cosmic rays from all directions can excite atmospheric molecules. Chemical reactions between some gases present in the high atmosphere are also occurring. All of these processes result in the production of weak light energy. Although air glow is uniform across the sky, it is most readily observed toward the horizon because one is looking through a greater depth of the atmosphere, and therefore, the light is concentrated along those sight lines.

Chapter 5: Winds

Pressure differences in the atmosphere set the air in motion, and temperature differences generally induce pressure differences in the horizontal. In the atmosphere, other forces come into play on air parcels as well.

The rotation of the earth causes all parcels to be deflected. This apparent force is known as the Coriolis force (see p. 18). In the northern hemisphere, air parcels appear to be forced to the right; south of the equator, they are deflected to the left. This apparent force is strongest at the equator and decreases as you approach the north and south poles. In the absence of other forces, the Coriolis force causes the winds to blow along lines of constant pressure. In the northern hemisphere, winds blow in a counterclockwise direction around centres of low pressure and clockwise around centres of high pressure.

> **The pessimist complains about the wind, the optimist expects it to change; the realist adjusts the sails.**
> —William Arthur Ward

Any moving air parcel is also subjected to frictional forces, the most significant of which is the roughness of the

underlying land surface. Frictional effects, which deflect winds by as much as 40 degrees toward lower pressure, are negligible above 400 metres over flat land or water surfaces and above 700 metres over rough terrain. Over water, the deflection is rarely more than 10 degrees. When looking at wind plots on maps of pressure at sea level, the drag of friction results in an apparent spiralling of winds toward the low pressure centres. The opposite is true for high pressure systems. Here friction appears to cause an outward spiralling of air away from the centre.

Curvature in the flow results in another force that affects the movement of air. This force opposes the motion of air along a curved path. It is generally a slight effect that results in the movement of air outward and away from the centre of low or high pressure.

The aforementioned forces modify large-scale circulation and wind patterns. Geographical features such as mountains, valleys and large lakes have a multitude of local effects on wind patterns. In addition, small-scale differences in heating are responsible for several wind patterns.

Typhon, or Typhoeus, was the personification of intense windstorms, so a typhoon is a fierce storm.

Lake Breezes

Solar radiation is absorbed at different rates by land and water surfaces. Under sunny skies, air over land heats more quickly than air over an adjacent water surface. As daytime heating progresses, the difference between the air temperatures creates a corresponding pressure difference. The air over land has a lower pressure, and air over the water surface is pushed toward land by higher air pressure over the lake. The heated air rises, establishing a local circulation pattern known as a lake breeze circulation, as shown in Figure 5-1. The effect is most pronounced adjacent to large lakes, such as Williston and Atlin in the north and Okanagan and Kootenay in the south, but it is present around all lakes. This circulation can also have a noticeable effect on the presence of

Lake Breeze Circulation

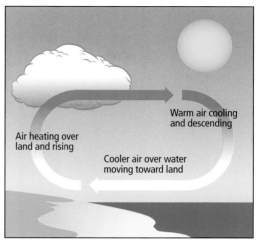

Warm air cooling and descending

Air heating over land and rising

Cooler air over water moving toward land

Fig. 5-1

clouds. Over land, the rising air parcels may have enough moisture to condense and form small cumulus clouds; subsiding air in the return flow over the lake is generally cloud free. And the circulation will progress during the day, strengthening and moving farther inland. The flow pattern may also become visible if smoke is present and becomes entrained in the air circulation.

Land Breeze Circulation

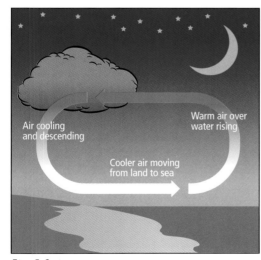

Air cooling and descending

Warm air over water rising

Cooler air moving from land to sea

Fig. 5-2

Land Breezes

The opposite effect occurs at night when solar heating has ceased. The land surface cools more quickly than the adjacent lake surface, and a point may be reached where surface pressures are higher over land than over the lake. Air is pushed offshore, establishing what is known as a land breeze circulation. This circulation is usually not as pronounced as the lake breeze.

Terrain Effect Winds

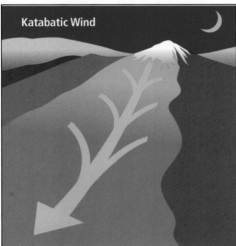

Fig. 5-3 Owing to heating and cooling, high terrain can substantially modify wind patterns.

Terrain Effects

High terrain can substantially modify wind patterns, often distorting what may otherwise be a uniform pattern based on the larger scale pressure pattern. During daylight hours, the sunlit side of a mountain may be strongly heated, and adjacent air will be significantly warmer than air farther out from the mountain. The warmer air moves up the slope to higher elevations while the air away from the mountain subsides. This movement is called the anabatic wind pattern. At night, a reverse circulation is established. In this case, the mountain surface radiates heat and cools the adjacent air more quickly than the air farther out from the mountain slope. The cool air begins to sink and flows down the mountain surface, producing what is called a katabatic wind. This wind can be quite pronounced if the mountain surface is covered with snow and cools adjacent air quickly. Katabatic winds can occur day or night in winter. They are usually much more pronounced than anabatic winds, and the mountain breezes that result can be quite cool. These thermally driven, circulations that are established in valleys and near mountains are often observed in British Columbia.

The trade winds get their name from the period in history when sailing ships were the main vehicles of trade. Captains of these ships factored in these winds when mapping out their journeys, sailing with the wind as much as possible.

Jet Streams

Large temperature differences in the atmosphere give rise to strong winds called jets. There are two common types of jets: the jet streams in the high atmosphere, and the low level nocturnal jets that appear in the lower atmosphere.

The jet stream is located near the top of the troposphere, about 10 kilometres above the earth's surface. Jet streams at British Columbia's latitude are usually associated with the frontal surface between air masses near the tropopause. At high altitudes, there may be several air masses, and therefore several jet streams, evident at any one time. Because the

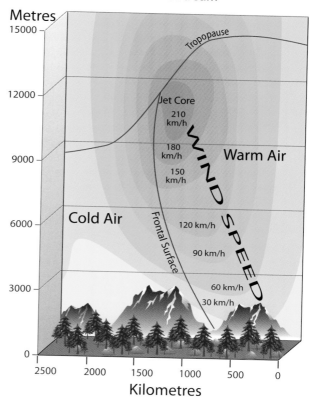

The Jet Stream

Fig. 5-4 Wind speed changes quickly above and below the jet stream core.

strength of the jet stream is a function of the temperature differences between air masses, the polar jet near the front that separates warm tropical air from colder polar maritime air is usually the strongest. And because colder air is poleward of warmer air, the wind within the jet stream usually blows from west to east.

Jet streams are weather phenomena with large linear dimension. When weather maps covering the whole hemisphere are examined, the jet streams may be seen to encircle the entire hemisphere.

Where vigorous weather systems are present in the lower atmosphere, the jet stream appears to break into segments as short as 1000 kilometres, but often a jet stream extends to 10,000 kilometres. The width of a jet stream band can range from 400 to 1500 kilometres.

Jet stream wind speeds are usually stronger in winter than in summer because of the greater contrast between warm and cold air masses. The jet stream band will vary greatly in width. The wind speed peaks at the core of the jet and can

Jet Stream Positions

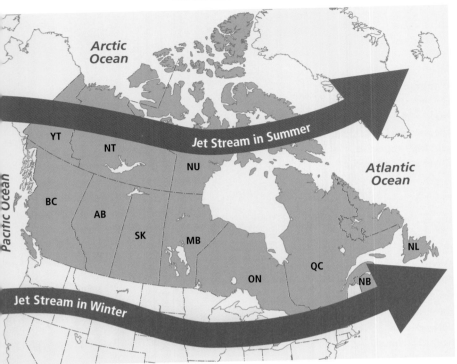

Fig. 5-5 The jet stream's average position moves north in summer.

reach speeds of over 400 kilometres per hour, but the maximums are generally in the 150 to 200 kilometre per hour range. The mean position of the jet stream also moves with the seasons. In winter, the strongest jet stream is usually positioned south of BC. In summer, the strongest jet stream cores may be located through or even north of BC for short periods.

For most of us, the importance of the jet stream is how it relates to air travel. Aircraft flying in the same direction as the jet stream have their speed enhanced by the same amount as the wind at flight level. So an aircraft flying across the continent from west to east may have a flight time that is 1½ hours shorter than an equivalent flight from east to west. Usually then, airlines prefer to fly on routes that take the jet stream into account.

The downside of flying the jet stream is the turbulence that is often associated with it. Turbulent flow in the atmosphere occurs when winds change speed rapidly. Jet stream flow is not horizontally uniform, and the speed drops more quickly poleward of the jet maximum than equatorward. As well, the wind drops quite quickly into the stratosphere above the jet core. So flying the jet stream is a balancing act between travel time and turbulence, and commercial airline pilots are constantly monitoring the locations of jet streams and associated areas of clear air turbulence.

Remember, the storm is a good opportunity for the pine and the cypress to show their strength and their flexibility.
—Ho Chi Minh

Nocturnal Jets

At night as the ground surface cools, so does the adjacent air. The air above the surface layer does not cool as quickly, so a warmer layer of air may be above the colder surface air. This situation is reversed from the normal change of temperature with height, in which air cools with elevation. The reversed lapse rate condition is known as an inversion. Nocturnal inversion conditions occur most nights. This thermal structure in the lower atmosphere often induces weakening of the wind adjacent to the ground surface, so the winds we experience at ground level are light overnight. At the top of the nocturnal inversion, the winds reach a maximum, which is known as a nocturnal jet. Maximum wind speeds can reach over 60 kilometres per hour. The depth of this nocturnal jet is only 100 metres or so, but because the nocturnal inversion is widespread and quite uniform, the nocturnal jet may exist as a "sheet" of strong winds several hundreds of kilometres wide and up to 1000 kilometres long. This jet breaks up in the morning as the nocturnal inversion dissipates with heating. On occasion, the jet becomes so intense overnight that the

Winds from Around the World

Brickfielder: a strong, dry, dust-laden summer wind from the desert in southern Australia.

Haboob: a strong wind that occurs in the Middle East and along the southern edges of the Sahara in the Sudan; is associated with extreme sandstorms and, occasionally, thunderstorms, and is most common in summer.

Levanter: a strong easterly wind in the Mediterranean, especially the Strait of Gibraltar; is associated with foggy, overcast or rainy weather, especially in winter.

Nor'easter: a strong northeast wind that blows across the Atlantic provinces and the coast of New England.

Sou'easter: a strong wind from the southeast that comes ahead of a Pacific storm system.

Pampero: a strong west or southwest wind in southern Argentina that brings with it cold polar air and is often associated with severe thunderstorms.

Santa Ana: a strong, hot, dry wind that blows from the southern California desert through the Santa Ana Pass; most common in spring and autumn, but can form at any time.

Sirrocco: a hot wind in southern Spain that originates in the Sahara Desert.

Squamish: a strong and often violent wind occurring in many of the fjords, inlets and valleys of British Columbia.

Taku: a powerful northeasterly wind that occurs occasionally in Alaska between October and April; winds speeds can reach hurricane levels.

stability of the atmosphere breaks down. In this case, surface winds may become gusty under clear sky conditions, usually accompanied by a temperature increase as the warm air aloft mixes to the surface. Usually the gusty conditions are not prolonged and the lower atmosphere repeats the inversion formation process. I have observed this phenomenon while working the fire weather forecast desk. After checking the overnight winds at some of the higher location weather stations, winds suddenly increased during the night for no apparent reason (i.e., there are no fronts or thunderstorms in the area).

Convection-Induced Winds

Cumulus clouds are formed when air near the ground is heated and begins to rise in features known as thermals. As the warm parcels rise, other air parcels move downward to replace the rising air. These parcels mix with the air at lower levels, resulting in a turbulent layer of air below the cloud base. An observer on the ground would feel the movement of this turbulent air as a gusty wind.

As cumulus clouds develop vertically, precipitation may start to form within

Wind is a major factor in spreading forest fires.

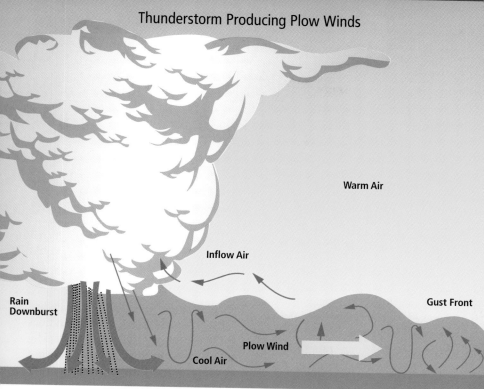

Fig. 5-6 Downward-rushing cold air moves rapidly along the ground, producing a plow wind close to the base of the storm and a gust front.

the cloud. The raindrops start to fall relative to the uplifting air, and as the drops grow, their downward velocity increases until it is faster than the upward moving air. This rain falls through the cloud, and when it exits the saturated base, evaporation starts. The evaporation process extracts heat from the surrounding air and may cause cooling. The air within the rain shaft may, in fact, become colder than surrounding air, causing an accelerated downward push. The downward rushing rain mass also drags the air immediately adjacent to the droplets so that an extensive slug of cold, moist air accelerates toward the ground. Upon reaching the underlying land surface, this downdraft spreads out laterally, pushing dryer air ahead. At the ground surface, we observe this movement as a gust front. The gust front may be located well downwind of the precipitating shower cell, and it can appear in any direction from the cloud. When a gust front associated with a precipitating rain cloud passes, there is sometimes a sudden wind shift with no precipitation. Most often, however, the wind shift is soon followed by rain. This gust front has a number of impacts. The cold air can inhibit further convection near the raining parent cloud by choking off

updrafts. The gust front can also undercut warm, moist air that is moving toward the parent cloud. Occasionally, the uplifted, moist air at the gust front starts a new line of convective showers. Intersecting gust fronts from thunderstorms with vigorous rain can result in the formation of a new rain cell, which may in turn grow into a mature thunderstorm.

The downburst from an actively precipitating thunderstorm occasionally results in strong winds that may damage trees or structures in their path. Winds in severe downbursts can exceed speeds of 250 kilometres per hour, the same speeds that are observed in tornadoes. In British Columbia, these downbursts can be seen in the form of "blowdowns" in forests throughout the province. The sudden violent winds can flatten trees, making the landscape look much as though a giant scythe had descended from the heavens and mowed the trees down. These winds can sometimes be confused with tornadoes, but if the downed trees are viewed from above, they line up in a linear fashion, showing no rotation. A blowdown that occurred in Manning Park back in the 1970s could be seen for many years next to the highway.

Downbursts are also a major hazard for aircraft during take-offs and landings because of the strong gusts and rapidly changing wind directions. Turbulence and shear can even trouble large aircraft and have been identified as causes for accidents.

Tornadoes

The winds associated with tornadoes are the strongest in the atmosphere. Conventional wind measurement instrumentation cannot withstand the force of these winds, so the strength is usually an estimate based on the type of damage that occurs. The speed scale is named after Theodore Fujita, the research scientist who developed it in the 1960s. The so-called F-scale describes wind speeds in six ranges, from F0, with wind speeds up to 115 kilometres per hour, to F5 tornadoes, with wind speeds estimated over 415 kilometres per hour. In BC, few tornadoes have speeds above the F1 level;

however, on rare occasions F2 and F3 tornadoes have been documented.

The tornado formation process is still not fully understood, but a credible theory is emerging based on observations of the tornado storm environment together with the understanding of thunderstorm dynamics. The majority of thunderstorms go through a normal lifecycle, from formation to decline, within about one hour. A few cells, known as supercells, last for several hours and can track hundreds of kilometres. Some supercells develop into what are called mesocyclones, with which the stronger tornadoes are associated.

Updrafts and downdrafts are associated with thunderstorms. As the thunderstorm cell moves forward, low level winds feed moisture into the cell below its base. This moist air condenses, releases latent heat and creates air parcels that can accelerate upward in areas known as updrafts. The speed in the updrafts can be more than 30 metres per second. Downdrafts composed of downrushing, cold, precipitation-laden air can have speeds in excess of 10 metres per second. These updrafts and downdrafts are often adjacent to each other, and the zone between them is subjected to a high wind shear. This wind shear can impart a horizontal rolling action to

Fujita Scale Rating the Severity of Tornadoes

The Fujita scale is used to rate the severity of tornadoes as a measure of the damage they cause.

INTENSITY	ESTIMATED WIND SPEED	DAMAGE
F0	light winds of 64–116 km/h	some damage to chimneys, TV antennas, roof shingles, trees, signs and windows
F1	moderate winds of 117–180 km/h	cars overturned, carports destroyed and trees uprooted
F2	considerable winds of 181–252 km/h	sheds and outbuildings demolished, roofs blown off homes, and mobile homes overturned
F3	severe winds of 253–330 km/h	exterior walls and roofs blown off homes, metal buildings collapsed or severely damaged, and forests and farmland flattened
F4	devastating winds of 331–417 km/h	few walls, if any, left standing in well-built homes; large steel and concrete missiles thrown great distances
F5	incredible winds of 418–509 km/h	homes levelled or carried great distances, tremendous damage to large structures such as schools and motels, and exterior walls and roofs can be torn off

Fig. 5-7

the atmosphere just above the ground, near the cloud base. This rolling wind tube sometimes becomes vertically tilted, and if the rotation is cyclonic (counter-clockwise), it may then impart a cyclonic circulation to the thunderstorm, creating a mesocyclone. As the mesocyclone continues its movement, ingesting warm, moist air and generating downdrafts, roll clouds continue to form and may be tilted onto a vertical axis. These rotating tubes stretch upward with the updraft, and if they have a counter-clockwise

> *If you are outside when a tornado touches down, the safest place to be is in a ditch.*

rotation, they may link with the larger mesocyclone. The stretching process further enhances the rotation speed of the tubes, much like how figure skaters increase their rotation speed by pulling

their arms closer to their body. At this point, the spinning column becomes a tornado. Approximately 15 to 20 percent of supercells produce tornadoes. The most severe tornadoes often have smaller spinning vortices that move around the main rotating funnel. These are known as suction vortices, and they frequently contain the strongest winds.

The length of time a tornado is on the ground is highly variable. The parent storm complex often generates a series of tornadoes.

Tornadoes are quite rare in British Columbia, averaging only one or two per year. The Central Interior Plateau has the most, with some F2 or F3 strengths observed. Some severe thunderstorms develop in the Southern Interior each year and are accompanied by strong winds, heavy downpours and hail. A small number of these cells likely grow even stronger, producing supercell thunderstorms, and 15 to 20 percent of supercells produce tornadoes. Weak tornadoes have probably occurred over the years in sparsely populated areas.

Formation of a Tornado

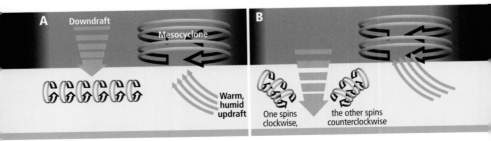

Fig. 5-8
A - low-level wind shear forms a spinning tube of air
B - the storm's downdraft pushes down on the spinning tube, tilting it into two columns

Tornado damage

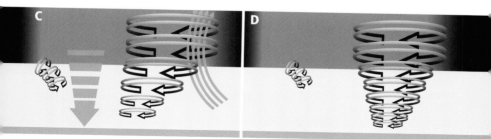

C - the counter-clockwise column is stretched into a tornado by the updraft
D - the column spinning clockwise, unstretched, spins slower. Sometimes it circles the
tornado as a small funnel cloud.

95

Dust Devils

Dust devils are basically weak cousins of tornadoes. They are not associated with any cloud and are formed during late spring and summer on nearly calm, sunny days by the heating caused by intense isolation on a dry or sandy terrain. Air in contact with such surfaces can become super heated. Hot air rises rapidly in a column above the hottest area of the terrain, and hot air swirls in at the base of the column, so that the whole column takes on either a clockwise or counter-clockwise rotation. Dust devils are made visible by the dust, sand or debris that they pick up from the ground. The most vigorous dust devils can extend to heights approaching one kilometre, though most are only a few metres in depth. Surface wind strength can be up to 70 kilometres per hour. Dust devils are usually short lived and move slowly in the direction of the mean surface wind.

Cold air funnels can hang below intense convective rain clouds.

> Fanatics in power and the funnel of a tornado have this in common—the narrow path in which they move is marked by violence and destruction.
>
> —Oscar Ostlund

A dust devil over Vernon (opposite page)

Cold Air Funnels and Waterspouts

Cold air funnels are small funnel clouds that rarely reach the earth's surface. When they do touch ground, and by definition become tornadoes, they are short lived. Their strength is usually in the F0 category, and they cause little damage, if any. These weather phenomena are not common in BC and mostly

Windstorms and Tornadoes

Tornado Watches

- Issued where there is significant potential for cold core funnels, waterspouts or rotating thunderstorms that may spawn tornados
- Can be issued for an individual region or a group of regions
- Lead time up to 2 hours, but usually much shorter
- There is time for preparation (secure loose items outdoors, notify others)
- Inform public to watch the sky for developments

Tornado Warnings

- Issued when a tornado is expected to develop soon, a tornado is nearby and will move into the area or a tornado is in the area
- Specific to a region or an area within a region
- Lead time up to 10 minutes but can be shorter
- Take cover

Windstorm and Tornado Safety

Outdoors

- Do not chase storms or tornadoes—they can change direction unexpectedly
- Get out of the storm path by moving at right angles to the direction of the storm's motion
- Get out of mobile homes or campers and go to a sturdy, permanent building
- If driving, special care and attention is required
  - obey traffic laws
  - if you stop, do not stay in the car
  - do not seek shelter under a road overpass

Indoors

- Go to the lower floors, such as a basement under stairs
- Put as many walls as possible between you and outdoors
- Avoid windows, especially large ones
- Washrooms, hallways, stairwells and reinforced walls are best
- Avoid large open-span roofed areas such as gyms
- Hide under heavy desks, tables or workbenches

Fig. 5-9

Tornado damage

occur in winter when cold arctic air pours out of the inlets and flows over the relatively warm sea surface. A line of increasingly larger convective cells can be seen on satellite images moving away from land. These cells often bring snowsqualls to the eastern shore of Vancouver Island. Sometimes, a small funnel will briefly form, and if it reaches ground, a very weak tornado results. The funnels usually occur over water and thus are called waterspouts. They aren't strong enough to inflict significant damage but could overturn a small boat if it was unlucky enough to be in one's path.

In Greek mythology, Zephyrus was the personification of the west wind, so a zephyr is a gentle breeze.

Chinook arch

Chinooks

These winds are formed along the lee slopes of mountain ranges with strong west to southwest winds aloft. They are the best known of the mountain-induced circulations. The Chinooks east of the Coast Mountains aren't as common or dramatic as the ones east of the Rockies, but they can still cause sudden wind shifts and rises in temperature. Chinooks occur when Pacific air warms as it descends along the eastern slopes of the mountains, scouring out the colder arctic air that lies in the lee of the Coast Mountains and the Rockies. The Chinook (from the Salish Indian word *tsinuk*, for "snow eater") is marked by the sudden onset of strong winds and rapidly warming temperatures. Because the air mass in the down-rushing air has lost most of its moisture during the climb from near sea level on the west coast, Chinook winds are very dry and have a desiccating effect. They can be moderately strong and gusty in British Columbia, with speeds often in excess of 70 kilometres per hour. When Chinook winds are funneled through narrow valleys aligned from east to west, even stronger winds can occur, occasionally over 100 kilometres per hour. These winds can be strong enough to force vehicles off the road.

The higher frequency of Chinooks and the strong westerly winds in the Peace River country make the area around Chetwynd an excellent location for wind turbines. Four 27-metre-high windmills have been installed in the region to light up the trees lining one side of the highway that runs through town. Twenty-five trees with a total of 5200 LED lights provide illumination year-round.

A new wind farm called the Dokie Project is in the works 40 kilometres northwest of Chetwynd. When it is completed (by autumn 2009), eight giant turbines will produce enough energy to power 34,000 homes a year, which would be enough to meet the needs of the city of Prince George and then some!

The Formation of Chinook Winds

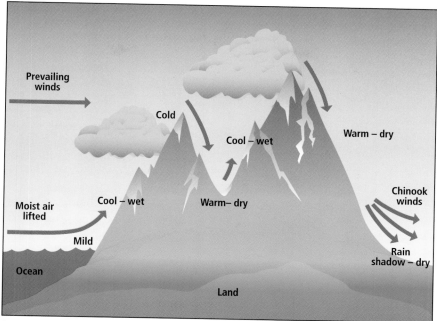

Fig. 5-10 Moist air is forced to rise over a succession of mountain ranges. The Chinook winds sweep down the lee of the Rockies.

The Tower of Winds in Athens

Built in 100 BC, the tower is an eight-sided structure, with each side depicting one of the main winds recognized by the ancient Greeks. The Greeks personified the winds, giving them personalities and attire that represented the type of weather each one brought. The eight winds recognized by the Greeks were Boreus (north), Notos (south), Zephyros (west), Apeliotes (east), Kaikas (northeast), Euros (southeast), Lips (southwest) and Skiros (northwest).

Chapter 6: Optics and Acoustics

The atmosphere consists of a mixture of gases, which have optical and acoustic properties and can produce illusions as we observe and listen. To understand the optical effects, it is first necessary to appreciate that the electromagnetic radiation emitted by the sun extends over a wide range or spectrum of wavelengths. The solar spectrum starts with the shortest and most energetic waves, known as gamma and x-rays, and then extends into the ultraviolet wavelengths. Visible light is in the middle of the spectrum, and infrared radiation, which we sense as heat, has the longest wavelengths.

Blue Sky

Why is the sky blue? This question was answered in the 19th century when scientists realized that gas molecules scatter electromagnetic radiation. As sunlight enters the top of the atmosphere, the gases immediately start scattering some waves in the light spectrum. The size of the gas molecules determines which wavelengths are scattered. The atmosphere consists of mostly oxygen and nitrogen, and the molecules of these gases are the right size to scatter blue light while leaving the longer wavelengths basically untouched. The blue light rays bounce off other molecules as well. This scattering occurs in all directions—back toward the sun, at right angles and even at small angles to the solar rays—and gives the sky its apparent blue colour from the horizon to the sun. If one were up at the top of the atmosphere where there is little oxygen or nitrogen, the sky above would appear black—there is nothing to scatter light. As one progresses toward the earth's surface, the atmosphere takes on a dark blue colour as some blue light is scattered. Indeed, at the tops of mountains, the sky generally has a darker blue colour.

Rainbows

Falling water droplets characteristically have a spherical shape. When a light ray passes through a water droplet, it reflects off the back of the inside surface and then passes back out the front. Each time the ray moves between the air and water surface, some refraction occurs and, like the action of a prism, the white light spreads into its colour spectrum. A rainbow is the observed action of millions of droplets acting together. The angles between the sun, the raindrops and the observer are all well defined by

Why the Sky is Blue

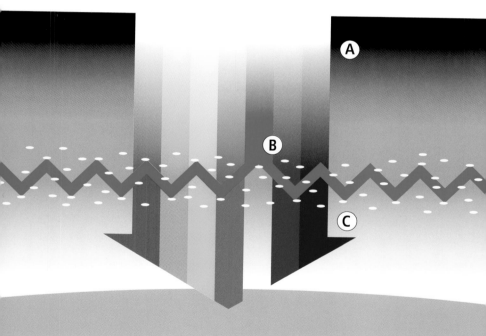

Fig. 6-1

A - white sunlight at the top of the atmosphere is composed of the full colour spectrum

B - some light in the blue colour spectrum is scattered in all directions by gas molecules in the atmosphere. As a result, we see blue sky in all directions.

C - the remaining light in other frequencies travels to the bottom of the atmosphere optical theory with the result that the rainbow is essentially an arc that appears when the sun is at the observer's back, shining on a mass of raindrops. So, the rainbow appears in the opposite direction of the sun when the sun is relatively low in the sky and is illuminating a falling mass of raindrops. If we were to observe a bank of raindrops from an airplane or mountaintop, we would see a complete circle.

What has been described above is known as the primary rainbow—the result of one reflection off the back of the rain droplet. Because the droplets are spherical, in reality multiple reflections occur inside each drop, giving rise to the appearance of a second or even third rainbow. With each reflection and refraction, there is some loss of brightness or intensity. As a result, the secondary bow is significantly less bright than the primary bow. This secondary bow is located outside the primary bow, and the colour sequence is reversed. The tertiary bow, when observed, appears as an extremely faint arc inside the primary bow.

In theory, multiple bows are possible, but because of increasing loss of intensity, they are not observed.

> **Old weather saying:**
> *Rainbow in the morning gives you fair warning.*

Formation of Primary and Secondary Rainbows

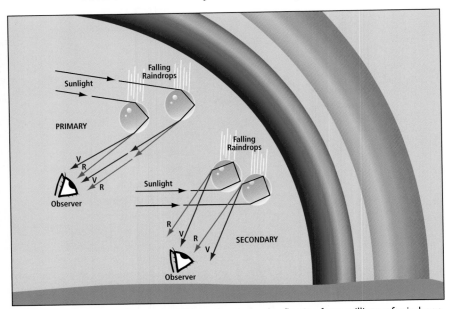

Fig. 6-2 Rainbows are the result of light refracting and reflecting from millions of raindrops.

Primary and secondary rainbows

In the secondary rainbow, the colour scheme is reversed; instead of having red on the outside and blue on the inside, as do primary rainbows, the outside is blue and the inside is red.

Sundog in Manning Park.

Sundog Formation

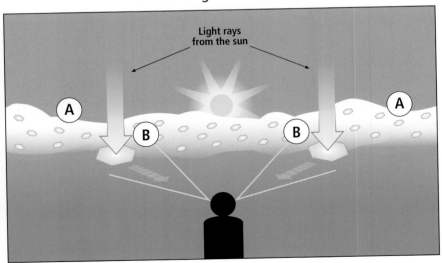

Fig. 6-3

A - cloud contains flat ice crystals

B - sun enters crystals and is bent at a 22 degree angle toward observer. Sundogs form
   on both sides of sun

Ice Crystal Phenomena

Several phenomena observed in British Columbia's winter skies are a function of the unique characteristics of atmospheric ice crystals. The shape of the water molecule determines the shape of the resulting ice crystals. When formed in calm wind conditions, ice crystals are often six sided, or hexagonal. Sunlight can reflect off the ice crystal surfaces and can also pass through the crystal, which splits the white light in a prism-like fashion. This bending and splitting of light is known as refraction. Two commonly observed winter phenomena result: sundogs and light pillars.

Sundogs or, more correctly, parhelia are observed when the sun passes through a layer of uniform ice crystals. Refraction of the sun's rays at the air-ice surface of each ice crystal produces the characteristic prismatic colour band. When the ice crystals lay uniformly in a horizontal plane, the refracted light is concentrated enough to form the sundogs, which can be seen at an angle of 22 degrees on either side of the sun. On cold winter days, a layer of ice crystals may form near ground level. These ice crystals are often called diamond dust because they reflect sunlight, giving the impression of glittering diamonds. On cold, calm winter mornings, the combination of diamond dust and ice crystals aloft gives the most spectacular displays of sundogs.

Light Pillar from Sunlight

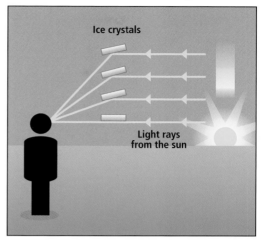

Fig. 6-4
Light reflected off underside of ice crystals results in a light pillar extending above the sun near the horizon

Light Pillar from Street Light

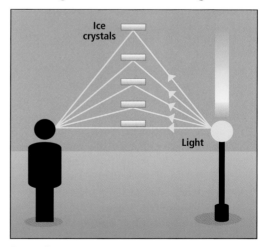

Fig. 6-5
Light reflected off ice crystals results in a light pillar extending above a street light

107

Light pillar extending above the sun

A **light pillar** is a reflection phenomenon that is formed in the presence of diamond dust. The more or less uniformly oriented ice crystals act individually as tiny reflecting surfaces. As shown in Figures 6-4 and 6-5, the resulting effect is a shaft of light that seems to extend above and below the light source. Any relatively intense light source will produce light pillars. During cold, calm nights, streetlights often appear to have pillars extending vertically above them.

Mirages

A mirage is a refraction phenomenon that occurs when light passes between layers of air with different densities. Light rays are bent as they pass from colder air to intensely heated air close to the earth's surface. Such conditions happen when the sun heats sand or asphalt surfaces. Temperatures in the lowest metre or so above such surfaces can often vary up to 10° C, resulting in a significant air density difference between

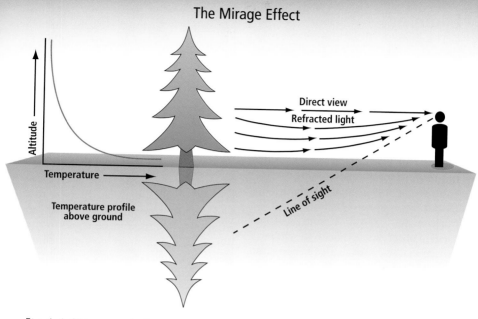

Fig. 6-6 Objects on the horizon may appear inverted.

immediately adjacent layers of air. The light rays passing between the air layers bend, not in a pronounced way, but when we look toward the horizon, we often see an intense blue image, which is in reality the bending of light from above the horizon. Occasionally it is not the sky that we see but rather the image of an object on the horizon. In this case, the image is turned upside down and is usually distorted. Mirages are generally unstable because the extreme vertical temperature differences also cause the warmer, lighter air parcels at ground level to move upward and be replaced by denser, cooler air immediately above. The result is turbulence in the mirage layer, resulting in an unstable, shimmering effect.

A mirage

Looming

This optical effect is similar to the mirage except that light rays are bent toward rather than away from the horizon. The condition happens when warm air is located above a colder layer near the earth's surface, which frequently occurs in the morning when the air adjacent to the colder ground is overlain by warmer air aloft. Because this event occurs most mornings, looming is a relatively frequent occurrence, usually manifested by an apparent tilting up of the horizon. As a result, at sunrise it can seem to observers as though they are in a broad, shallow bowl. This phenomenon also affects shorter wavelengths, which affects the behaviour of weather radar beams. Radars emit microwave radiation, which behaves much like light waves do. Weather radars emit their microwave beam at a slight angle above the horizon. However, during strong inversion conditions, the radar rays are bent toward the ground, even beyond the true horizon. The ground reflects these rays back along the same path through the atmosphere to the radar receiver. As a result, the radar display shows an image of ground clutter for some distance around the radar's location. In effect, the radar is "seeing" the bottom of the shallow bowl.

Other Optical Effects

Because much of weather observation depends on watching the sky, people throughout the centuries realized that they could interpret what the weather would be like in the short range by observing sky and cloud conditions. Much of this observation became folklore, but some of it has an element of scientific rationale. An often-heard expression is "red sky at night, sailor's delight; red sky at morning, sailor take warning." The explanation for this saying comes, in part, from the atmosphere's light-scattering capabilities, and from the direction of storm motion at mid-latitudes. At sunrise and sunset, the sun's rays are travelling through a long pathway, and the residual light is mostly red, giving cloud surfaces a reddish tint. At sunset, if there is no cloud in the western sky, clouds illuminated to the east look red. At mid-latitudes, storms move from west to east, and no cloud to the west indicates there are no approaching storms. So, "red sky at night [i.e., at sunset], sailor's delight" means no storms are on their way. If the sky is red in the morning, the rising sun is not obscured by cloud, and it may be illuminating the underside of a deck of cloud to the west. Again, because storms move from the west to east, it means potential storm clouds are approaching. Therefore, "red sky at morning [i.e., at sunrise], sailor take warning" means a storm is on its way. So, though it is folklore, the old saying has some basis in scientific reality.

Acoustical Effects

Like light waves, sound waves propagate through the atmosphere, and they have some similar behavioural properties that are notable under inversion conditions. The most frequently observed effect is that of reflection at discontinuities in the atmosphere. Sound waves are bent or refracted at relatively sharp discontinuities in the vertical thermal structure. At night, nocturnal temperature inversions form when air in contact with the earth's surface cools more

Refraction of Sound Waves

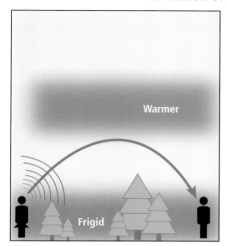

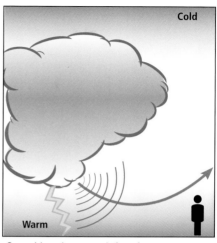

Fig. 6-7 Sound bends downward in a temperature inversion

Sound bends upward if surface is warm

quickly than the air aloft, resulting in the reverse of the normal situation in which temperature drops off slowly with height. In this inversion situation, there can be a sharp discontinuity in temperature. Sound waves propagating at an angle toward this discontinuity are refracted downward toward the earth's surface. This phenomenon is most readily apparent when the observer is relatively distant from a steady source of noise, such as roadway with heavy traffic. At night, the noise from a distant highway, train or even general traffic noise from a city seems to be louder than it is during the day. Of course, other factors such as background noise levels and wind direction also come into play, but the effect can be quite noticeable—for example when one normally does not hear a freeway, but it becomes quite notable after sunset when the inversion becomes well established. Sound waves can also be bent upward when0 temperatures decrease rapidly with height. For

example, we often do not hear thunder from a distant lightning stroke because the sound wave is bent upward and away from us.

A simple acoustical effect relates to the rate of propagation of sound waves in air. Sound waves propagate at a speed of approximately 330 metres per second in the lower atmosphere. Light, on the other hand, is transmitted essentially instantaneously. We can use this difference to calculate a rough estimate of the distance to a lightning stroke by counting the number of seconds between seeing a lightning flash and hearing the crack of thunder. A pause of three seconds means the sound travelled about 1 kilometre, six seconds indicates 2 kilometres and so on. The timing of distant lightning strokes becomes more complicated because of reflection of the sound off atmospheric discontinuities, such as edges of clouds and shafts of falling rain.

111

Chapter 7:
British Columbia's Climate

Controls and Influences

Many people confuse the terms weather and climate and use them interchangeably. The two are closely related; the main difference is a matter of time scale—weather lasts a few days to several months; climate spans decades. If you add up all the temperature and precipitation values at a location over the year, you have performed the first step toward determining the climate. Extend these records over a number of years and take the average, and you will see patterns start to emerge. There are cooler and drier years as well as warmer and wetter

ones, but the average gives you a general idea of what a "normal" year will bring.

Average conditions don't give you a complete picture of an area's climate; there are two other factors to examine— the standard deviation from the mean, and extreme readings. The standard deviation from the mean is very important. For example, consider the average annual snowfall in Victoria; the figure in the climate records shows 44 centimetres, but there is a huge deviation from this number. In many years, little if any snow falls in Victoria; however, there

BRITISH COLUMBIA CLIMATE EXTREMES 1971–2000
(Adapted from the Meteorological Service of Canada data
as posted on the Environment Canada website)

CHILLIWACK
MAXIMUM (°C)   37.8 ON JULY 27, 1958
MINIMUM (°C)   -21.7 ON DECEMBER 29, 1968
DAILY RAINFALL (MM)   122.6 ON DECEMBER 13, 1979
DAILY SNOWFALL (CM)   66.8 ON NOVEMBER 16, 1996
SNOW DEPTH (CM)   55.0 ON JANUARY 9, 1991

CRANBROOK
MAXIMUM (°C)   36.6 ON JULY 21, 1985
MINIMUM (°C)   -40.0 ON DECEMBER 30, 1968
DAILY RAINFALL (MM)   53.1 ON MAY 22, 1968
DAILY SNOWFALL (CM)   30.2 ON DECEMBER 29, 1996
SNOW DEPTH (CM)   80.0 ON DECEMBER 30, 1996

DAWSON CREEK
MAXIMUM (°C)   34.5 ON AUGUST 9, 1981
MINIMUM (°C)   -49.2 ON DECEMBER 29, 1992
DAILY RAINFALL (MM)   75.2 ON JUNE 11, 1990
DAILY SNOWFALL (CM)   35.2 ON MAY 13, 1986
SNOW DEPTH (CM)   86.0 ON FEBRUARY 6, 1982

FORT ST. JOHN
MAXIMUM (°C)   33.6 ON AUGUST 9, 1981
MINIMUM (°C)   -47.2 ON JANUARY 30, 1947
DAILY RAINFALL (MM)   80.3 ON JUNE 27, 1965
DAILY SNOWFALL (CM)   47.8 ON MAY 22, 1960
SNOW DEPTH (CM)   112.0 ON MARCH 28, 1974

KAMLOOPS
MAXIMUM (°C)   40.6 ON JULY 31, 1971
MINIMUM (°C)   -37.2 ON JANUARY 29, 1969
DAILY RAINFALL (MM)   48.0 ON AUGUST 16, 1976
DAILY SNOWFALL (CM)   33.8 ON JANUARY 7, 1962
SNOW DEPTH (CM)   81.0 ON DECEMBER 25, 1971

KELOWNA
MAXIMUM (°C)   39.5 ON JULY 24, 1994
MINIMUM (°C)   -36.1 ON DECEMBER 30, 1968
DAILY RAINFALL (MM)   33.8 ON JULY 21, 1997
DAILY SNOWFALL (CM)   26.0 ON JANUARY 23, 1982
SNOW DEPTH (CM)   56.0 ON FEBRUARY 20, 1975

NANAIMO
MAXIMUM (°C)   36.7 ON AUGUST 9, 1960
MINIMUM (°C)   -20.0 ON DECEMBER 30, 1968
DAILY RAINFALL (MM)   97.3 ON FEBRUARY 1, 1991
DAILY SNOWFALL (CM)   73.7 ON FEBRUARY 12, 1975
SNOW DEPTH (CM)   74.0 ON JANUARY 4, 1966

PRINCE GEORGE
MAXIMUM (°C)   36.0 ON MAY 29, 1983
MINIMUM (°C)   -50.0 ON JANUARY 2, 1950
DAILY RAINFALL (MM)   50.0 ON AUGUST 4, 1948
DAILY SNOWFALL (CM)   30.8 ON DECEMBER 30, 1990
SNOW DEPTH (CM)   104.0 ON FEBRUARY 24, 1956

PRINCE RUPERT
MAXIMUM (°C)   28.7 ON AUGUST 14, 1977
MINIMUM (°C)   -24.4 ON JANUARY 4, 1965
DAILY RAINFALL (MM)   118.2 ON SEPTEMBER 25, 1983
DAILY SNOWFALL (CM)   39.9 ON JANUARY 20, 1973
SNOW DEPTH (CM)   76.0 ON JANUARY 9, 1965

REVELSTOKE
MAXIMUM (°C)   37.2 ON JUNE 26, 1992
MINIMUM (°C)   -29.4 ON JANUARY 17, 1970
DAILY RAINFALL (MM)   43.6 ON SEPTEMBER 4, 1982
DAILY SNOWFALL (CM)   60.2 ON DECEMBER 10, 1980
SNOW DEPTH (CM)   173.0 ON FEBRUARY 15, 1982

VANCOUVER
MAXIMUM (°C)   33.3 ON AUGUST 9, 1960
MINIMUM (°C)   -17.8 ON JANUARY 14, 1950
DAILY RAINFALL (MM)   89.4 ON DECEMBER 25, 1972
DAILY SNOWFALL (CM)   41.0 ON DECEMBER 29, 1996
SNOW DEPTH (CM)   61.0 ON JANUARY 15, 1971

VICTORIA
MAXIMUM (°C)   36.1 ON JULY 16, 1941
MINIMUM (°C)   -15.6 ON JANUARY 28, 1950
DAILY RAINFALL (MM)   92.8 ON JANUARY 18, 1986
DAILY SNOWFALL (CM)   64.5 ON DECEMBER 29, 1996
SNOW DEPTH (CM)   67.0 ON DECEMBER 30, 1996

Fig. 7-1

have been some incredible snow years. In 1996–97, more than 140 centimetres were recorded throughout the winter, and 124 centimetres fell in the month of December alone! The city also had heavy snow events in 1916 and 1923.

Extreme readings show the entire range of possibility. In the temperate zone, the westerlies prevail, which generally results in close to average conditions. If the upper winds switch to a meridional flow (meaning from the south to north and vice versa), extreme temperatures—either hot or cold—and sometimes heavy precipitation events are the result.

The World Meteorological Organization (WMO) has set criteria in which 30 years of data are needed to determine a location's climate. After this length of time, there should be sufficient data to show both average and extreme values. The current set of values for BC runs from 1971 to 2000. These values can be accessed on the internet at http://www.climate.weatheroffice.ec.gc.ca/Welcome_e.html

Across British Columbia there is as much variety in climate as in the day-to-day weather. There are both maritime and continental influences, and when you add the topographical effects to the mix, you can see why so many microclimates exist in the province.

Air Masses

An air mass is defined as a large body of air, hundreds to thousands of kilometres across, with relatively similar temperature and humidity values. The assortment of air masses that move through BC show characteristics of the regions where they were formed. Most originate over the Pacific, and the type of weather they bring is determined mainly by the latitude they stream from.

In the colder seasons, northwest flows often bring maritime arctic air masses to BC. The word arctic is included in the name because the colder land areas of Alaska and the Yukon usually influence the air mass before it is modified over the north Pacific. Typically the weather is cooler than average with a showery type of precipitation. Ski resorts really like this air mass because freezing levels are at a fairly low altitude, bringing significant snowfall to the mountains.

Except in winter, the most common air mass that lies over the province is the maritime polar air mass. The source region for this air lies over the mid-latitude Pacific Ocean. The prevailing westerlies bring a succession of high pressure ridges and low pressure troughs. Through spring and autumn, the storm track is farther south, so rainy periods are more frequent along the coast, while the Interior gets showers or flurries. In the summer months, the jet stream shifts northward, bringing much drier conditions to southern portions of BC. The northern half of the province has its warm and dry periods, but the storm track is never far away. Occasional frontal systems still bring periods of rain to the north coast, with scattered showers or thundershowers progressing inland. Temperatures are usually close to average or slightly above with maritime polar air masses.

Winter Air Masses and Circulation

Continental Arctic
- very cold, -25 to -50° C
- dry, very stable
- pronounced temperature inversion

Maritime Arctic
- very unstable
- clouds, frequent showers or flurries
- visibility good except in showers

Maritime Polar
- milder and more stable than arctic air

Pacific Maritime Tropical
- light winds, cooler than Atlantic air
- comes to North America from west or northwest
- stable in lower 1000 m (marine stratum)

Atlantic Maritime Tropical
- comes to North America from south or southeast
- warm and humid

SST - Sea surface temperature

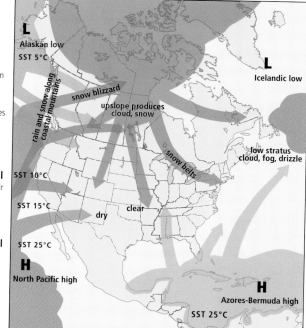

Summer Air Masses and Circulation

Continental Tropical
- hot, dry, unstable

Maritime Arctic
- continental air modified by open seas, lakes and swamps

Maritime Polar
- warmer and more stable than maritime arctic air

Pacific Maritime Tropical
- high pressure blocks moist air

Atlantic Maritime Tropical
- oppressively hot and humid
- unstable, frequent thunderstorms

SST - Sea surface temperature

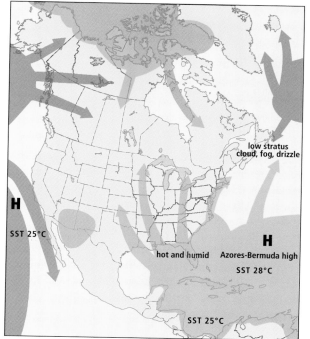

Fig. 7-2

A shift to the southwest in the upper flow brings a warming trend to the province because this air originates over water with much higher sea surface temperatures. This type of air mass is called a Pacific maritime tropical air mass, and it can at times be responsible for dramatic weather events. Once or twice each winter, a frontal zone sets up from just north of the Hawaiian Islands, aimed at the south coast of BC. This "pineapple express" can bring extremely heavy rainfalls to the coast, with wet snow spreading inland. Floods often develop in coastal areas, and with high elevation freezing levels, the avalanche danger rises to extreme over the mountains in the Interior. In summer, the air mass is usually much drier, and the "Hawaiian High" brings extended periods of sunshine and very warm temperatures to southern BC (and, if the ridge is strong enough, to the entire province). If the flow switches more to the south, convective bands of cloud bring a threat of thunderstorms.

From December through February, with the storm track well to the south, relatively clear skies prevail across the Yukon and the Northwest Territories. The nights are long, so outgoing radiation hits a peak and temperatures plummet. The air mass that forms under these conditions is called a continental arctic air mass. The bitterly cold air often moves southward to cover northern sections of BC, southward to the Peace, with only periodic blasts of maritime air—i.e., Chinooks—scouring out the cold air. When the upper flow bends around to the north or northeast several times per winter, the arctic front drops southward. The cold air travels relatively unimpeded across the Interior Plateau until higher terrain, especially the Coast Mountains, acts as a barrier to its passage. When the cold air piles up, it seeps through the mountain passes and coastal inlets until it reaches the Lower Mainland and eventually Vancouver Island and the Queen Charlottes. By this time the temperatures have moderated considerably, but the strong outflow winds can still result in high wind chills in exposed locations. Continental arctic air is usually quite dry, but if there is any interaction with moist maritime air, heavy snowstorms can develop.

Marine Influences

British Columbia has close to 23,000 kilometres of coastline (9000 along the mainland coast and 14,000 around islands), so the Pacific Ocean and inland waterways have a tremendous influence on BC's coastal weather. In summer months as temperatures warm over land surfaces, pressures drop, inducing a daily sea breeze effect. The result is cooler temperatures along the adjacent coastline and often, to a lesser degree, for many kilometres inland. There is a silver lining, though. In winter, the relatively warm ocean waters bring much milder temperatures to the coast compared to the sub-freezing readings in the Interior. Early in my weather career, I worked at Cape St. James, an isolated lighthouse on the southern tip of the Queen Charlotte Islands. Seasonal temperature changes were minimal, ranging from an average high temperature in August of 17° C to a low of 3°C in January—an annual difference of 14° C. A Chinook wind in Alberta can change the temperature by as much in less than an hour!

Sea surface temperatures are fairly uniform from winter to summer because water has a large capacity to absorb and store heat but is a poor conductor. The high sun angle and increasing hours of summer sunshine gradually warm the waters along the coast. However, it is well past the June summer solstice and into August when the warmest water readings are measured. Correspondingly, the end of July into early August is climatologically the warmest time of year on the coast.

From October through to the end of March, the storm track from the Pacific can be aimed at BC for days or even weeks at a time. Weather forecasters use the phrase "a series of frontal systems" to avoid being more specific about the timing and ending of each disturbance as it approaches and moves over the coast. To keep it brief, the forecast simply states "periods of rain today and tomorrow." There are usually sunny breaks between systems, but the post-frontal showers often let up for only a few hours before the next frontal wave brings more steady rain. From April to September, frontal systems are weaker, and with the storm track mostly lying farther north, rainfall amounts are much less. However, even with more sunshine, sea breezes still bring fairly high humidity levels, and fog banks commonly lie along or just off the outer coast.

Pacific Ocean Currents

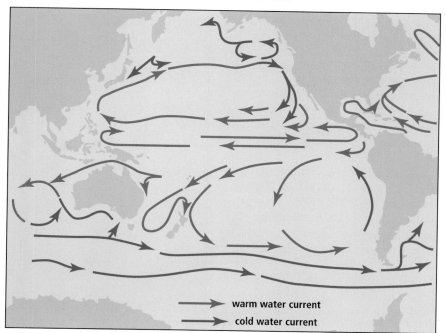

warm water current
cold water current

Fig. 7-3

The marine influence farther inland is less pronounced but still significant. By inland, I am referring to areas away from the water, such as Terrace and Bella Coola in the north, the east and south coasts of Vancouver Island and the Lower Mainland eastward through the Fraser Valley to Hope. In these areas, there is less temperature and moisture modification, but as long as westerly winds prevail, a coastal type of climate is experienced. If the flow is strong enough, frontal systems spread rain up the coastal inlets and through the passes into the Interior. We used to have an expression for this situation at the weather office in Kelowna when steady rain was falling in the normally dry Fraser Canyon— "Lytton is going coastal!"

Terrain Effects

Each mountain range in British Columbia has an effect on frontal systems as they move inland from the Pacific. My first introduction to this phenomenon was tuning into the nightly weather spot on CBUT-TV in Vancouver way back in the 1960s. Bob Fortune, long-time weather presenter and host of the outdoor show Klahanie, would often use his chalkboard to draw in the mountains throughout BC. He would start with the Insular Mountains on Vancouver Island and the Queen Charlotte Islands, then move eastward to the Coast/Skeena and Columbia to Cassiar ranges and finally to the Rockies. He would then explain the current weather situation and how each rise and fall across higher terrain modified the weather as the disturbance moved across the province.

A close look at a map of southern BC shows numerous branches extending from the main mountain ranges. Any traveller driving from the Okanagan through to Alberta is very aware of this as he or she climbs through the passes of the Monashee, Selkirk, Purcell and Rocky mountains before finally reaching flatter ground. The terrain in northwestern BC is even more complex, with geography texts showing dozens of small ranges and individual mountains. The only regions in the province where relatively flat ground is the norm are the northeast and the Interior Plateau.

As moist air reaches the coast, it is forced upward, and as it cools and condenses, rain intensifies. As a result, the western slopes of Vancouver Island and the Queen Charlotte Islands receive the highest annual rainfall amounts in North America. The air warms and

BC's mountainous terrain has a huge effect on our climate.

dries as it descends to the east coasts of the islands, reducing rain amounts as a "rain shadow" develops. The shadow effect is very pronounced on the south coast of Vancouver Island. In southwesterly flows, air warms and dries as it descends from the 2500-metre slopes of the Olympic Mountains in Washington State. In many winter storms, Victoria receives little if any rain, while a short distance away heavy amounts are experienced. The annual precipitation on the Victoria waterfront is 600 millimetres; 80 kilometres to the west, Port Renfrew records 3600 millimetres!

As the air continues eastward, it rises, cools and precipitates on the west-facing slopes of the Coast Mountains, then once again dries out as it sinks into the Interior Plateau. Because this air already lost much of its moisture during its previous descent on the inner coast, it brings very low annual rainfalls to areas just to the east of the mountains, such as the Chilcotin, Nicola and Similkameen. Driving up the Coquihalla Highway clearly demonstrates the upslope/downslope effect. Often it can be foggy and raining from Hope until you reach the Coquihalla summit; then abruptly the rain eases (or ends entirely), and sometimes the sun even peeks out as you make your way down the other side. The contrast in vegetation is dramatic—from lush coastal rainforest on the way up from Hope to sagebrush and cured grass by the time you arrive in Merritt in the Nicola Valley. The same rising and falling process continues eastward across the rest of the province. More rain and snow are deposited on windward slopes than on lee slopes or in the valleys.

Cold Lows

Cold lows or upper lows are nearly circular weather features in which temperatures, both surface and aloft, become cooler toward the centre of the low. A significant difference between cold lows and regular low pressure centres is that cold lows are not embedded in the upper flow circulation. Bands of cloud and showers rotate around the low, sometimes hundreds of kilometres distant from the centre.

Computer models have always had a hard time predicting the path and speed of these lows, often showing them moving too fast. They can travel through the province at any time of year but are more frequent in spring and early summer. If you look at the climate records in BC, many locations show an increase in precipitation amounts from late May until early July. As a low moves inland, areas to the north experience an easterly flow. When this happens in the foothills of the Coast Mountains and the Rockies, a moist upslope effect can bring heavy rainfalls to normally dry areas. The Chilcotin and Peace River country are especially prone to this wet pattern.

Because the air mass close to the low is unseasonably cool, freezing levels are at low altitudes, and higher mountain passes can receive wet snow even in the middle of summer. The cool, unsettled, showery weather can continue for several days if the low slows down. When the low finally exits the province, quite often there is only a brief period of warming and drying before the next low moves inland from the Pacific. For some unknown reason, cold lows seem to target the May long weekend to arrive.

Later in summer, convective bands ahead of the low can move northward and northeastward through the province, creating numerous lightning strikes and fire starts. Fortunately, as the low finally moves inland, showers and cooler temperatures help dampen the intensity of the fires.

El Niño / La Niña

The global atmospheric circulation pattern undergoes periodic changes, and El Niño of the tropical Pacific Ocean has a significant influence on the weather in British Columbia as well as in other parts of the world. This phenomenon demonstrates the link between circulation patterns in the atmosphere and those in the earth's oceans. The main currents of the Pacific basin are shown in Figure 7-3.

Typical January–March Weather Anomalies and Atmospheric Circulation During Moderate to Strong El Niño & La Niña

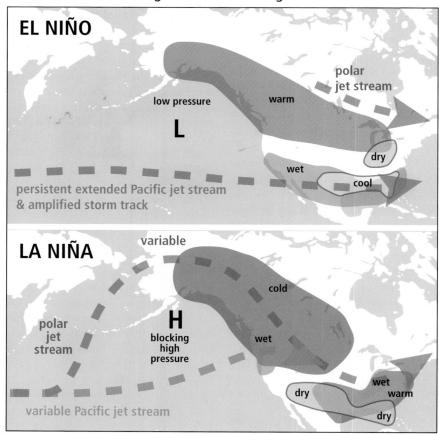

Fig. 7-4

The trade winds that blow from east to west along the equatorial Pacific drag surface ocean waters toward the west. Accompanying the westward drift of water is an upwelling of cool, deep, nutrient-rich water along the South American coast, which gives rise to a productive fishing industry off Peru and Ecuador. The trade winds occasionally slacken or even reverse, and the accumulated water off the Indonesian coast begins to shift back across the Pacific, taking about two months to reach South America. When the bulge of warm water reaches the South American coast, it cuts off the upwelling of the nutrient-rich waters. The warming event usually starts in December, and the local fishermen named it El Niño, Spanish for "the little boy" and a reference to the Christ child. When the waters in the eastern Pacific are colder than normal, what is known as La Niña (or "the little girl") occurs. La Niña is the antithesis of El Niño.

El Niño warming events occur on an irregular basis at intervals of two to seven years. Accompanying this oceanic cycle is an oscillation in atmospheric pressure on either side of the Pacific Ocean. At the onset of El Niño, average pressures over the western Pacific are higher than normal and are accompanied by lower than normal pressures in the east. This see-saw cycle is known as the Southern Oscillation. The ocean and atmospheric cycles are linked and are known as the ENSO or El Niño-Southern Oscillation.

Strong El Niño and La Niña episodes have a noticeable effect on the weather conditions in western North America. In winter during an El Niño event, a stronger area of low pressure lies well off the BC coast and a relatively stationary ridge of high pressure lingers over the province, often positioned near the Rockies. There are usually two storm tracks, one directed at the Alaska Panhandle and a southern one passing through California. So, during a strong El Niño event, winters in BC are milder and drier than average. During a strong La Niña, the ridge of high pressure lies over the Pacific, bringing a northwesterly flow over the province. The jet stream moves through Alaska before heading southeast toward the coast. Storms and cold air masses are guided by the jet stream, so winters are colder than average with somewhat higher precipitation amounts. A lot more snow accumulates in the mountains, and even coastal areas have more snow events.

During El Niño summers, the trend is still for below average precipitation and warmer than average temperatures. Two recent episodes, in 1998 and 2003, resulted in extremely dry conditions in the Southern Interior. Numerous wildfires destroyed hundreds of homes and thousands of hectares of forest. The 2003 fire season was the province's most costly on record. La Niña events are most pronounced in winter, but there are some trends in other months as well. Precipitation amounts tend to be above average while temperatures are near to slightly below normal. During the most recent event, in 2007–08, most of the province experienced a very late spring. Temperatures were considerably below average until early June. Freezing levels remained low, and snow continued to fall in the mountains with some late-season killing frosts in the valleys.

Temperature

Temperature is the weather element that has the greatest influence on living things. Temperature variations largely control the viability of plants, and temperature means and extremes define human comfort levels. In British Columbia, the temperature pattern across the province is an indication of the variation of the sun's angle from the south to the north. There are also major effects from marine and topographical features that produce large differences in temperature at locations only a few kilometres apart horizontally or a couple of thousand metres apart vertically.

The temperature variation through the year on the boreal plains and forests of northeastern BC is a reflection of the continental climate. Winters are long and cold, and summers are short and relatively cool. Spring and autumn months can see wide swings in temperature over short periods as cold arctic and warmer marine air masses constantly shift north and south.

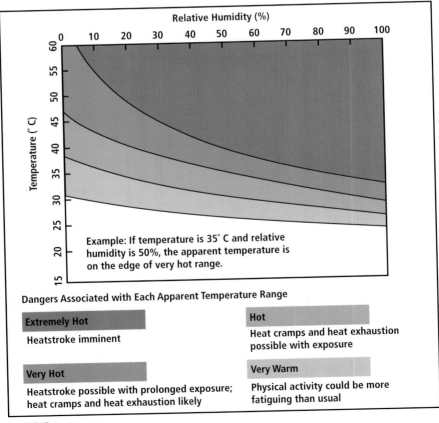

Apparent Temperature

Example: If temperature is 35° C and relative humidity is 50%, the apparent temperature is on the edge of very hot range.

Dangers Associated with Each Apparent Temperature Range

Extremely Hot
Heatstroke imminent

Hot
Heat cramps and heat exhaustion possible with exposure

Very Hot
Heatstroke possible with prolonged exposure; heat cramps and heat exhaustion likely

Very Warm
Physical activity could be more fatiguing than usual

Fig. 7-5

Over the Interior Plateau, northern regions are dominated by a continental climate in winter with temperatures closely resembling the cold values of those to the east of the Rockies. To the south where the Pacific storm track prevails, temperatures are quite a bit warmer with the marine influence. In summer, the sun's angle has the biggest effect, with areas in the north staying cooler than areas in the south. However, though the north is still cooler than the south in summer, the difference between the regions is less than in winter because of the longer days up north.

Temperatures are modified year-round on the coast, as is typical of a marine climate. There is little daily temperature contrast, and summer season maximums and winter minimums show a lot less separation than over the interior regions of the province.

The average lapse rate is about 6° C per 1000 metres. Therefore, mountain regions of the province are considerably cooler throughout the year. Occasionally an inversion in this pattern occurs, commonly within a strong ridge of high pressure. The temperature rises with height for a couple of thousand metres, then starts to cool above the inversion.

Extreme Temperatures

The hottest temperatures in the province are in the lower elevations of the Southern Interior and the Fraser Canyon. There is long-standing debate about which town holds the extreme maximum temperature record; Osoyoos, Lytton and Lillooet all vie for the top position. Looking at the records, Lillooet and Lytton share the second highest maximum temperatures in Canada with

44.4° C recorded on July 16, 1941. Osoyoos is close behind with 42.8° C on July 27, 1998. However, it is likely that Osoyoos should be the true record holder for the hottest all time BC temperature; when temperature readings from Osoyoos are compared with those from Lytton and Lillooet, the Osoyoos temperatures are usually the highest, but records there unfortunately don't go back as far as 1941.

Smith River, just south of the Yukon border, has the dubious honour of being the coldest spot in the province. On January 31, 1947, the temperature bottomed out at -58.9° C. Another cold spot is Puntzi Mountain on the Chilcotin Plateau. With an extremely cold arctic air mass over the area, -52.8° C was recorded on December 29, 1968. Residents of the area who were brave enough to go outside and risk almost instant frostbite reported that the sound of the trees cracking as the sap froze up was deafening!

Urban Heat Islands

Cities have a modifying effect and create local microclimates known as urban heat islands. The most apparent effect is on temperature because buildings, factories and moving vehicles emit large amounts of waste heat, and concrete buildings, asphalt parking lots and roadways absorb heat during daylight hours and release it overnight. The larger the urban area, the more pronounced the heat island. Vancouver and denser populated areas of the Lower Mainland have the only significant heat island effect in the province. However, Victoria, Kelowna, Kamloops and Prince George all have some degree of warming.

Average January Daily Temperature

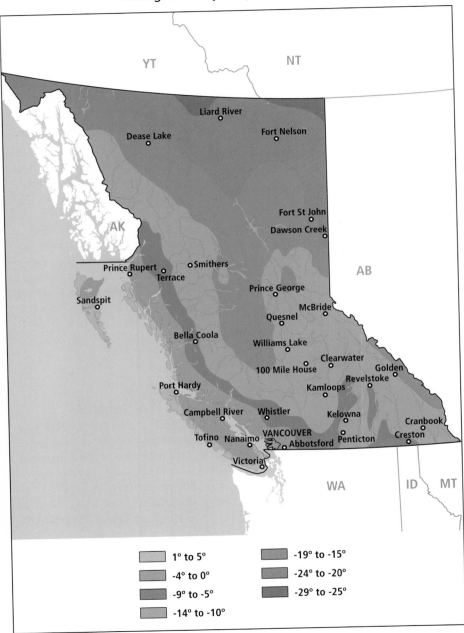

Fig. 7-6

Average July Daily Temperature

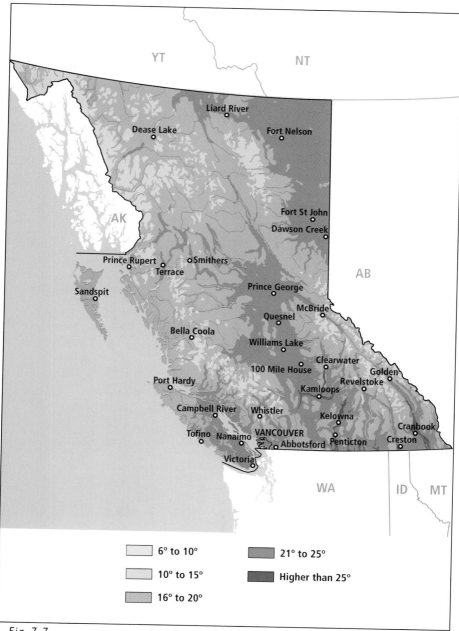

6° to 10°	21° to 25°
10° to 15°	Higher than 25°
16° to 20°	

Fig. 7-7

Heating Degree Days

A practical application of temperature statistics is the calculation of the Heating Degree Day (HDD). Heating Degree Days (see Figure 7-8) are directly related to the amount of energy required to heat buildings to a comfortable level. Heating is generally required when external temperatures are below some reference value, usually 18° C. The Heating Degree Day is calculated by subtracting the average daily temperature from 18. For example, if the average temperature for a day is 10° C, then the HDD value would be 8; or if the day's average is -10° C, then the HDD value is 28. Daily temperatures above 18° C have an HDD of zero. The HDD values can be summed over a year to give a useful estimate of heating energy requirements. The BC coast and Southern Interior have lower HDD values, with a general rise toward northern BC.

Growing Degree Days

The summer growing season varies greatly across BC. The Growing Degree Day (GDD) is a useful measure of the amount of heat available to initiate and sustain crop growth. The GDD is related to how far the average daily temperature varies from a specific reference temperature. The daily values are accumulated over the growing season. The reference temperature depends on the type of plant, but 5° C is often used for general plant growth. Figure 7-11 shows the Growing Degree Days above 5° C

for BC. Most of Vancouver Island, the Lower Mainland and the Southern Interior valleys have the highest number of Growing Degree Days. As one might expect, the values decrease in the north, except for along the Rocky Mountain Trench and portions of the Peace.

What Are Heating Degree Days?

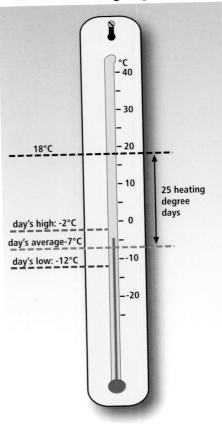

Fig. 7-8 The day's average temperature is subtracted from 18° C to get the heating degree days.

Frost Dates

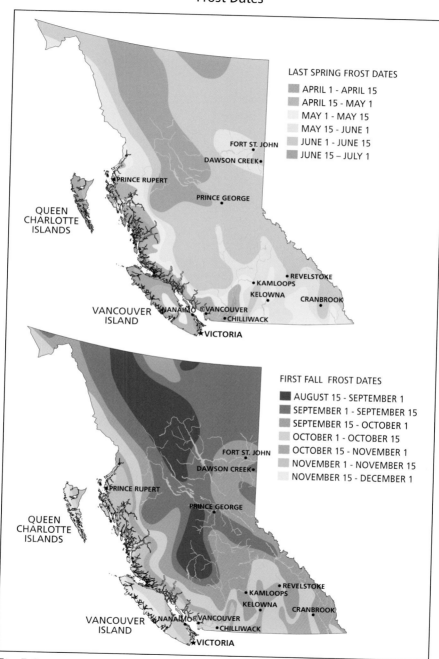

Fig. 7-9

Annual Total Heating Degree Days Below 18° C

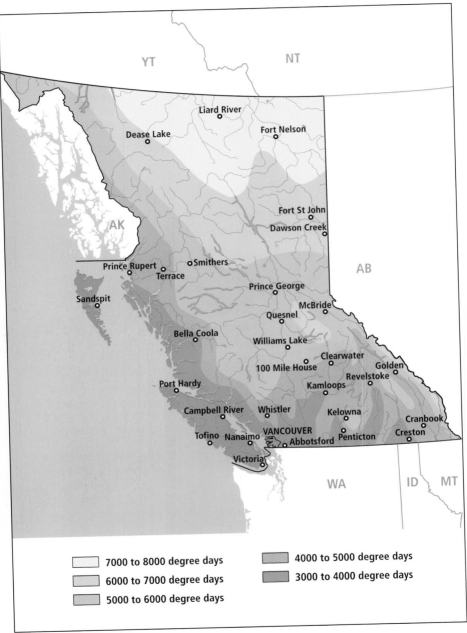

Legend:
- 7000 to 8000 degree days
- 6000 to 7000 degree days
- 5000 to 6000 degree days
- 4000 to 5000 degree days
- 3000 to 4000 degree days

Fig. 7-10

Annual Total Growing Degree Days Above 5° C

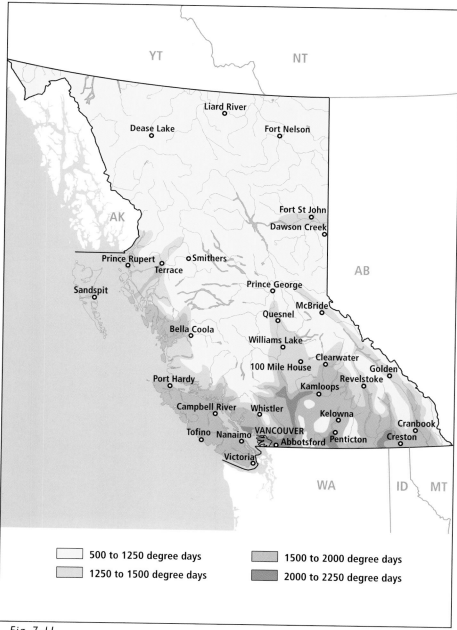

500 to 1250 degree days
1250 to 1500 degree days
1500 to 2000 degree days
2000 to 2250 degree days

Fig. 7-11

Wind Chill

Wind chill is a measure of how quickly heat is lost from exposed skin. The wind chill index has been developed to measure this cooling effect and to signal conditions that may be harmful. The wind chill calculation table provides "feels like" values in degrees Celsius, indicating how cold the temperature feels once the strength of the wind has been factored in. Wind chill values can fall below -30° C at most Northern Interior locations each winter. In the south, such cold readings are reached occasionally, but not every winter. Wind chill is only a factor on the coast when cold outflow winds develop during arctic outbreaks.

Extreme Wind Chills at some BC Weather Stations

Station	Wind Chill	Date
Fort Nelson	-55.6	Feb. 3, 1968
Williams Lake	-52.2	Dec. 15, 1964
Prince George	-51.5	Jan. 24, 1972
Smithers	-50.1	Dec. 16, 1964
Terrace	-42.2	Dec. 31, 1968
Penticton	-39.7	Dec. 29, 1968
Vancouver	-27.8	Dec. 16,1964
Victoria	-25.1	Dec. 16, 1964
Nanaimo	-22.4	Jan. 25, 1969
Tofino	-17.9	Jan. 30, 1969

Fig. 7-12

Wind Chill Calculation Table

V_{10} \ T air	5	0	-5	-10	-15	-20	-25	-30	-35	-40	-45	-50
5	4	-2	-7	-13	-19	-24	-30	-36	-41	-47	-53	-58
10	3	-3	-9	-15	-21	-27	-33	-39	-45	-51	-57	-63
15	2	-4	-11	-17	-23	-29	-35	-41	-48	-54	-60	-66
20	1	-5	-12	-18	-24	-30	-37	-43	-49	-56	-62	-68
25	1	-6	-12	-19	-25	-32	-38	-44	-51	-57	-64	-70
30	0	-6	-13	-20	-26	-33	-39	-46	-52	-59	-65	-72
35	0	-7	-14	-20	-27	-33	-40	-47	-53	-60	-66	-73
40	-1	-7	-14	-21	-27	-34	-41	-48	-54	-61	-68	-74
45	-1	-8	-15	-21	-28	-35	-42	-48	-55	-62	-69	-75
50	-1	-8	-15	-22	-29	-35	-42	-49	-56	-63	-69	-76
55	-2	-8	-15	-22	-29	-36	-43	-50	-57	-63	-70	-77
60	-2	-9	-16	-23	-30	-36	-43	-50	-57	-64	-71	-78
65	-2	-9	-16	-23	-30	-37	-44	-51	-58	-65	-72	-79
70	-2	-9	-16	-23	-30	-37	-44	-51	-58	-65	-72	-80
75	-3	-10	-17	-24	-31	-38	-45	-52	-59	-66	-73	-80
80	-3	-10	-17	-24	-31	-38	-45	-52	-60	-67	-74	-81

T air = air temperature in ° C and V_{10} = observed wind speed at 10 m elevation, in km/h.

Frostbite Guide

Low risk of frostbite for most people

Increasing risk of frostbite for most people in 10 to 30 minutes of exposure

High risk for most people in 5 to 10 minutes of exposure

High risk for most people in 2 to 5 minutes of exposure

High risk for most people in 2 minutes of exposure or less

Fig. 7-13

Humidex

The degree of discomfort humans experience in warm weather depends on the rate at which our bodies can lose heat through the evaporation of perspiration. This natural cooling mechanism is compromised when atmospheric humidity is high enough to inhibit evaporation. The humidex index (Figure 7-14) combines air temperature and relative humidity into a number that is a perceived temperature. When humidex values are higher than 30, some people experience discomfort, and when the value is over 40, everyone is uncomfortable. Because BC stays cool, much of the air on the coast and the summer Interior heat is on the dry side, so humidex values don't often exceed 35.

HUMIDEX (°C)	DEGREE OF DISCOMFORT
20 – 29	Comfortable
30 – 39	Varying degrees of discomfort
40 – 45	Almost everyone uncomfortable
46+	Active physical exertion must be avoided
54+	Heat stroke may be imminent

Fig. 7-14

Precipitation

Annual and seasonal precipitation amounts are highly variable across British Columbia. Southwest and west-facing slopes along the mountain ranges throughout the province receive the heaviest precipitation, and the driest areas lie to the east of the Coast Mountains. There is a significant difference in the amount of rain from winter to summer along the coast as winter storms give way to summer sunshine, but the Interior receives a fairly even distribution throughout the year. Precipitation is more frequent in winter, but it is light; it rains less often in summer, but the amount of rain that falls in the showers and thunderstorms is heavier.

In winter, frontal systems embedded in the Pacific storm track give the coast much more precipitation than the Interior. The outer coast is the wettest, while areas to the lee of higher terrain have much less precipitation. In the Central and Northern Interior, almost all precipitation falls as snow. To the south, the valleys have a mix of snow and rain, while higher elevations have almost all snow. Precipitation is much higher along and near the mountainous terrain of eastern BC, and it peaks to the east of Revelstoke in Glacier National Park. Conditions are much drier to the west near the Coast Range, as well as to the north of Prince George where arctic air with less moisture content prevails.

With the seasonal pattern of the storm track shifting northward in summer, rain amounts on the south coast are much lower. The east and south coasts of Vancouver Island often have very little rain all summer. The occasional frontal system still brings rain to the north coast but considerably less than in winter. The Interior shows a trend to higher values from west to east, and from south to north over the northern Interior Plateau. The area that receives the most rainfall, mainly because of the effects of upslope precipitation, lies in the Rockies to the west of Fort Nelson.

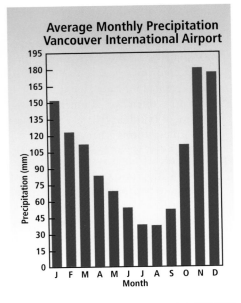

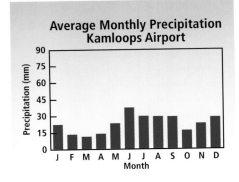

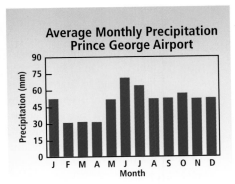

Precipitation Extremes

As previously mentioned, the western slopes of Vancouver Island record the highest rainfall amounts in the province. There is not much elevation gain from sea level, but it is enough to turn already heavy rain into a deluge. In 1997, more than 9300 millimetres were recorded, which isn't far behind the wettest spots on earth. Only Mount Wai-'ale-'ale on Kauai, one of the Hawaiian Islands, and a few locations in the monsoon belt in India are wetter.

Rain intensities can also be very heavy on the west coast of BC. On October 6, 1967, the Brynnor mine site (close to Ucluelet) measured 489 millimetres in just 24 hours—quite a bit more than some interior communities average in an entire year!

Ashcroft, to the lee of the Coast range, is considered the driest location in the province and in all of southern Canada. The average rainfall is a scant 250 millimetres. Sagebrush and ponderosa pine struggle to find moisture in this desert-like climate. The annual precipitation map (Figure 7-16) shows a wide swath from the Coast Mountains to the Columbia Mountains that receives less than 400 millimetres per year. This area stretches all the way from the U.S. border to north of Babine Lake in the Bulkley Valley. Other dry areas in the province are the Rocky Mountain Trench and the boreal plains east of the Rockies.

Fig. 7-15

Average Annual Precipitation

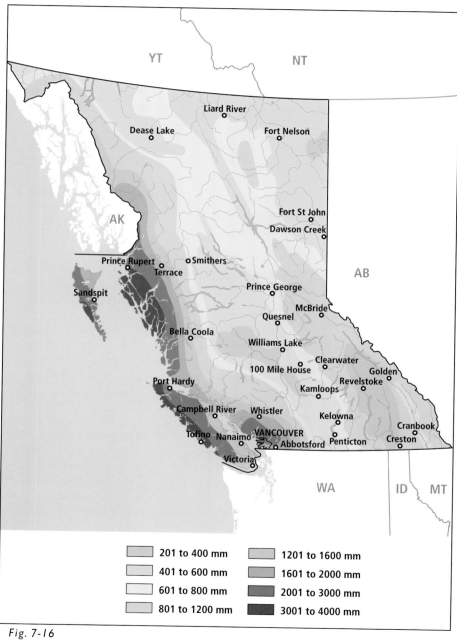

201 to 400 mm	1201 to 1600 mm
401 to 600 mm	1601 to 2000 mm
601 to 800 mm	2001 to 3000 mm
801 to 1200 mm	3001 to 4000 mm

Fig. 7-16

Snow

Figure 7-17 shows the yearly snow distribution across the province. Notice the heavier amounts on the higher elevations of Vancouver Island and the Coast, Columbia and Rocky mountain ranges. The Interior Plateau, northeastern BC and the Rocky Mountain Trench have considerably less accumulation, and lower elevations on the coast receive the least snowfall in most years.

The heaviest amounts in BC and all of Canada are in the Columbia and Rocky mountains. Only the Saint Elias Mountains in the southwestern Yukon likely receive more, but there are no measurements to confirm this information.

BC's mountains are prone to avalanches.

Avalanches

The huge amount of snow deposited each year on BC interior mountains is great for winter recreation. World-renowned waist-high powder and the thrill of travelling to remote parts of the province attract thousands of backcountry skiers and snowmobilers each year. The high-elevation snow is only a short trip by helicopter or the newer high-tech snowmobiles.

The pursuit of the ultimate adventure comes with some risks. Avalanches are always a threat, especially when the snow is unstable—the slightest disturbance is enough to trigger a slide. Over 80 percent of the time, people caught in avalanches survive, especially if they've taken training courses put on by the Canadian Avalanche Centre, based in Revelstoke. Unfortunately, some slides are so massive that there is no escape. An average of 14 people each year are killed by slides in Canada, mostly in southern BC in a triangle running from Vancouver Island to Pincher Creek in southern Alberta and up to Hinton in central Alberta.

A typical pattern that causes a high risk for avalanches starts with cold temperatures producing a slippery hoar frost layer on the snow. Then, when a flood of warm, moist air brings a period of heavy, wet snow, the weight of the top layer can easily slide on the icy surface below.

The winter of 2008–09 saw several of these cold to warm cycles, making the snow very unstable. More than 20 deaths were reported in BC through March 2009, 15 of which were snowmobilers. The most tragic accident occurred at the end of December 2008 when several avalanches buried a large group of snowmobilers near Fernie, in southeastern BC. Eight men, all from the nearby town of Sparwood, lost their lives.

Winter travel through mountain passes in BC can be hazardous, especially during storms. The snow maintenance companies do their best to keep the roads safe, but there are times when the snow just falls too quickly for the plows and sand trucks to keep up. There is also the ongoing threat of avalanches. In Rogers Pass, cutting through the Selkirk Mountains, Parks Canada employees attempt to keep the danger to a minimum by using the largest mobile avalanche-control program in the world. Five long tunnels through the pass lessen the chance of being trapped in a slide, but when the risk is highest, the Trans-Canada highway is closed until artillery fire or explosives dropped from helicopters bring down small slides to reduce the threat of a major avalanche. The avalanche control also protects Canadian Pacific Railway trains. Before this service and the construction of numerous snow sheds through the route, avalanches were common. In 1910, 62 people were killed in a single incident.

Sunshine

Figure 7-18 shows the average annual total hours of sunshine in BC. There is a noticeable rise in the hours from west to east, reflecting the change from the cloudier coast to the brighter Interior. The areas that receive the most sunshine are in the Thompson, Okanagan and Kootenay regions, where the sun shines for more than 2000 hours each year. The

The beach at Harrison Hot Springs

central and north coasts of BC are the dullest, with many locations averaging less than 1200 hours annually. Prince Rupert has the fewest hours of sun of any Canadian city. A South African family suffering from a rare genetic skin disorder that is aggravated by exposure to the sun moved to the city in 1999 after searching worldwide for a suitable location with the least amount of sunshine. The exception to the generally dull coast is southern Vancouver Island, where Victoria receives about 2050 hours annually because of subsidence and drying from the Olympic Mountains.

Average Annual Snowfall

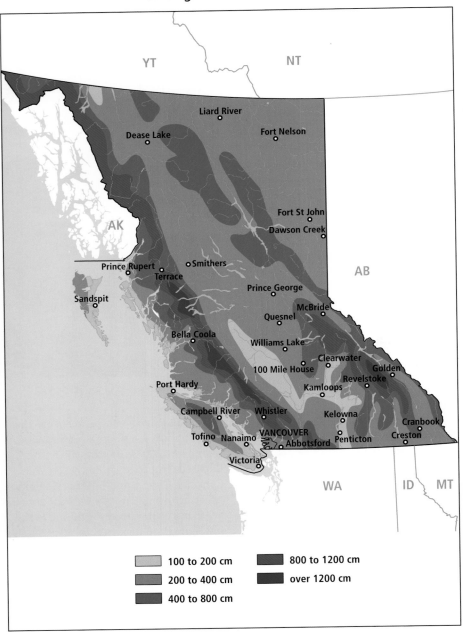

Fig. 7-17

Average Annual Total Hours of Sunshine

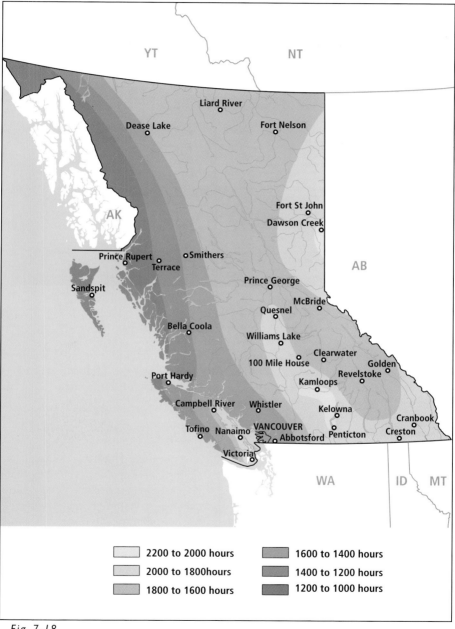

Fig. 7-18

2200 to 2000 hours

2000 to 1800 hours

1800 to 1600 hours

1600 to 1400 hours

1400 to 1200 hours

1200 to 1000 hours

Chapter 8: Storms

Depending on what part of the province you come from, the word storm is open to different interpretations. In Smithers, your first thought might be snowstorms; in Nelson, a violent thunderstorm might come to mind; in Tofino, most likely you'll picture a howling Sou'easter. That's not to say that summer storms never rage up north, or that heavy snow doesn't fall in the Kootenays or on the coast. Some types of storms are common, and others are quite rare.

Satellite images often show Pacific storms lined up offshore, each awaiting its turn to make landfall on the coast. Marine storms lead the list in the number of significant weather events during the year. Coast dwellers become jaded some winters as a seemingly never-ending parade of frontal systems brings rain and gales every second or third day.

Severe summer storms occur less frequently, but they have major impacts on people, property and livelihoods. They tend to be brief in nature, most only lasting 5 to 10 minutes, but the damage they cause can run into tens of millions of

> **If you spend your whole life waiting for the storm, you'll never enjoy the sunshine.**
>
> —Mark Twain

dollars. Sudden strong and gusty winds, heavy downpours, frequent lightning strikes, hail and even on rare occasions tornadoes accompany these storms. And of course, too much rain can cause flooding.

In most winters, there are two or three incursions of cold arctic air that plunge southward, eventually covering the entire province. Inevitably, after a few days, the prevailing westerlies take over again and start to push the cold air northward. The battleground between these air masses can get quite messy. The warm, moist Pacific air slides over the much colder arctic air near the ground. It is this frontal lift that occasionally brings heavy snowfalls throughout the southern part of the province.

Other less common but still note-worthy storms include freezing rain, blizzards and, the rarest of them all, remnants of typhoons that slam into the coast.

Typhoon Freda, October 1962

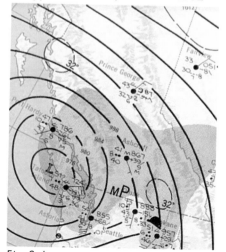

Fig. 8-1

Marine Storms

Technically, the storm in October 1962—which ranks as one of the most intense in south coast history—was not a typhoon. However, long-time residents still refer to it as Typhoon Freda (or the Columbus Day storm in Washington State). Some facts support the typhoon label—a minimum central pressure of 958 mb (equivalent to a category 3 hurricane on the Saffir-Simpson hurricane scale, Figure 8-2) and winds in excess of 180 kilometres per hour. The main elements of a typhoon were missing, though—the wind field was not as compact nor as strong as a tropical cyclone, and rainfall amounts were much less. The correct term for the storm was an extra tropical wave cyclone. I was living in Victoria at the time, and this storm had an incredible effect on me, assuring my life-long interest in weather. I can still recall the waves smashing into the break-water, with spray billowing over Dallas Road all the way into Ross Bay Cemetery.

The Pacific storm track brings the majority of low pressure systems to the coast during the mid-autumn through to the early spring months. Typically a low centre deepens as it approaches the north coast of BC then takes a northeast turn, ending up in the northern Gulf of Alaska. Frontal waves moving to the south of these lows bring periods of rain and strong winds to the north and central BC coasts. By the time the disturbances reach the south coast, they have started to weaken. Rainfall amounts and winds can still be intense on the outer coast, but once the front passes over Vancouver Island, winds have diminished, and the inner coast through to the Lower Mainland receives much less rain.

Saffir/Simpson Scale

Scale No.	Central Pressure		Wind		Damage
	MB	Inches	Mph	Knots	
1	>980	>28.94	74–95	64–83	Minimal
2	965–979	28.5–28.91	96–110	84–95	Moderate
3	945–964	27.91–28.47	111–130	96–113	Extensive
4	920–944	27.17–27.88	131–155	114–135	Extreme
5	<920	<27.17	155+	135+	Catastrophic

Fig. 8-2

These storms are not to be taken lightly, especially by ships in their path, but they are relatively weak by west coast standards.

A few times each winter, a low pressure centre approaching from the southwest tracks toward the north or central coasts of BC. As the low nears BC, the interaction of moist subtropical air with much cooler air to the north can cause a rapid deepening and intensification of the low. Weather forecasters call this explosive deepening of lows "bombs." Very strong winds and intense rains spread to the entire coast with these storms. Winds in excess of 100 kilometres per hour are common, rising to hurricane force in some exposed locations. Solander Island, located near Brooks Peninsula on northwest Vancouver Island, has recorded winds in excess of 160 kilometres per hour.

As the deep low centre and frontal system approach the coast, east to southeast, storm-force winds develop over waterways and exposed land areas. There is usually a brief lull after the low centre starts to move inland, then the winds veer and increase, becoming strong southwest to west winds in the strong pressure gradient behind the low.

December 2006 Storm

British Columbia had a very active autumn and winter storm season in 2006. With a constant flow of Pacific disturbances, November precipitation at the Vancouver Harbour climate station, near Stanley Park, measured 481 millimetres, more than double the average amount. There were only four dry days in the entire month! An invasion of arctic air brought cold and snow at the end of the month. Temperatures in Vancouver dropped to -12° C on November 28, and 38 centimetres of snow were measured from November 25 to 30. In the Fraser Valley, a record 44 centimetres of snow fell on November 26 at Abbotsford.

For a while in early December, the storm track shifted northward, with high pressure systems predominating and only light rainfalls recorded. This situation would all change beginning on December 11. A couple of fair sized storms ripped through the south coast, producing strong winds and heavy rains from December 11 through 13. Hydro crews were very busy maintaining power, and some BC ferries from Vancouver Island to the mainland were delayed or cancelled. On the morning of December 14, the computer models showed a rapidly

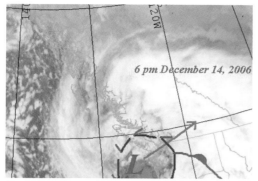

6 pm December 14, 2006

Fig. 8-3

southern Alberta, with closely spaced isobars along the BC south coast. Vancouver airport and Abbotsford reported westerly winds of 70 kilometres per hour with gusts over 100 kilometres per hour in the early hours of the morning. Some exposed locations had much stronger winds. Race Rocks, 10 kilometres from downtown Victoria, recorded 157 kilometres per hour.

deepening low pressure centre on a heading that would take it straight toward southern Vancouver Island. Environment Canada issued wind warnings for all land areas, and marine storm warnings were put in place for coastal waterways. Global-TV weatherman Mark Madryga warned viewers that this storm could be the most devastating of the season.

Easterly winds picked up, and the rain started in the Vancouver area around noon. The rain and gusty winds continued for the rest of the day. Figure 8-3 shows the satellite image at 6 OL on December 14. Notice the huge cloud shield extending all the way to the Central Interior and as far east as Saskatchewan. At this time, the occluded front (when a cold front overtakes a warm front) had just passed over Vancouver Island with the low pressure centre still lying offshore. The winds up to this point were fairly strong but were nothing out of the ordinary. As the low moved inland, a very strong pressure gradient developed behind it. The surface weather chart at 5 @L on December 15 showed the low already over

Damage was extensive, adding up to insured losses of $80 million. Downed power lines were everywhere, with a peak of 250,000 BC Hydro customers without service. Ferry schedules were disrupted, roads and schools were closed and thousands of trees were broken off or uprooted. Stanley Park, near downtown Vancouver, was hit exceptionally hard. More than 3000 trees (one in five) were lost, and the 9-kilometre sea wall walkway around the park was damaged so badly that it didn't reopen for almost a year. One of the theories offered to explain why such a large number of trees were downed is that the ground had been saturated from the recent heavy rains.

Trees down in Stanley Park, December 15, 2006

141

Top British Columbia Marine Storms

Location	Date	Anectodal
Southern BC	Apr. 13–14, 2002	wind caused whiteout conditions near Victoria; widespread damage and power outages from south coast to Salmon Arm area
BC Coast	Dec. 14, 2001	70–100 km/h northwest winds persisted over region for 12–18 hours; thousands of trees downed; ferries cancelled; widespread power outages
BC Coast	Oct. 22–23, 2001	remnants of a typhoon reached the coast; widespread damage and power outages
Vancouver	Mar. 3, 1999	60–90 km/h southwest winds; storm surge breached the seawall near Boundary Bay, flooding the area
Fraser Valley	Jan. 29, 1999	56 km/h winds with gusts up to 93 km/h; 2 mobile homes were blown off their supporting blocks and carried a meter or so by the wind
Victoria	Mar. 9, 1997	67 km/h winds; widespread power outages and downed trees
South Coast	Dec. 17, 1991	gale-force westerlies with gusts up to 85 km/h; widespread property damage, power outages and downed trees
Southern BC	Oct. 16, 1991	96 km/h gusts; 2 deaths; blowdown destroyed the equivalent of 6000 logging trucks of trees; widespread power failures and accidents
North Coast	Apr. 25–26, 1985	storm-force southeast winds; 3 fatalities and 7 fishing boats capsized at Hecate Strait
Vancouver Island	Apr. 15, 1984	southwest winds with gusts reaching 124 km/h; several million dollars in damages; extensive property damage and power outages from center of the island northward; capsized fishing and sailboats; winds fanned forest fires
Southwestern BC	Nov. 14, 1981	southeast winds with gusts exceeding 100 km/h; extensive power outages; freighters damaged in Vancouver
Victoria/ Vancouver	Oct. 13, 1962	Typhoon Freda; 145 km/h wind gusts in Victoria; 100 km/h gusts in Vancouver; 7 fatalities; $10 million in damages; 20% of trees in Stanley Park destroyed
Darling Creek	Apr. 30, 1943	gale-force winds; the Uzbekistan was driven ashore and sank; all crew members survived

Fig. 8-4

Summer Storms

From the end of May until early September, hardly a day goes by without at least a few thunderstorms developing across BC. With the stronger daytime heating creating the convection and lift necessary for cell growth, interior regions have significantly more storms. Lightning on the coast is much less common and is usually associated with sharp cold fronts or subtropical bands of convection invading from the south.

There is considerable debate as to what actually constitutes a thunderstorm. In fire weather forecasting, the term thunderstorm is usually reserved for cells that have no rain associated with them. The "dry lightning" that often develops with a high pressure ridge breakdown is a major cause of fires in BC. The air mass is so dry that most, if not all, the rain under the cloud base evaporates before it reaches the ground. The term thundershower is used when rain is expected with the cells.

The forecasts produced by Environment Canada use the term thunderstorm to indicate the potential for severe weather. Usually there is a severe thunderstorm watch or warning that goes along with the forecast. The term thundershower is used for the garden-variety type of storm when severe weather is not expected.

During the summer months, Environment Canada forecasters at the Pacific Storm Prediction Centre in Vancouver perform a detailed analysis each day to determine the potential for severe storms.

Here is Environment Canada's criteria for issuing Severe Thunderstorm watches and warnings in BC*:

Severe Thunderstorm Watch: issued when conditions are favourable for the development of severe thunderstorms with large hail, heavy rain, damaging winds or intense lightning.

Severe Thunderstorm Warning: issued when a severe thunderstorm has, or will develop and produce one or more of the following conditions:

Hail with a diameter of 10 mm or larger on the coast and 15 mm in the Interior. Heavy rain of 20 mm or more in one hour or less on the coast (25 mm in the Interior). Wind gusts of 90 km/h or greater. Intense lightning.

Severe Thunderstorm Warnings are also issued when a severe thunderstorm has the potential to produce a tornado.

*These criteria vary quite a bit across the country.

Dozens of factors affecting the moisture, air mass stability and wind patterns at various levels of the atmosphere are assessed to decide if there is a need to issue watches or warnings. On most days, there are regions in the province where thundershowers are likely to occur. However, there are only a few times each year when all the signs point to the potential of a severe thunderstorm outbreak.

Chilliwack Golf and Country Club

Flooding

There is long history of flooding along the Fraser River particularly in the Lower Fraser valley near Chilliwack. Residents had to deal with vast lakes covering their properties almost every year. Various projects on a limited scale were undertaken to hold back the flood waters. However, it wasn't until the devastating flood in June 1948, causing $20 million damage to property, that a major flood control dyke program was established.

The worst events are in the spring but even in winter, flooding has occurred. In early January 2009, warming temperatures melted the snow pack. Then a 'pineapple express' deposited over 100 millimetres of rain in a couple of days, causing widespread flooding adjacent to the Fraser.

Interior regions of the province are also vulnerable to floods. These are once again most likely in the spring but ice jams in winter can also cause damage along the Fraser, Skeena and Bulkley, as well as many other smaller rivers. For 66 days, from December 10, 2007 until February, 2008, the city of Prince George was in a constant state of alert. Ice jams on the Nechako and Fraser rivers flooded properties, sometimes with little advance warning.

Tornadoes

Tornado events are quite rare in BC, with an average of only one or two reported across the province each year. On the Fujita scale (see p. 93), most only rate F0 or F1.

However, very strong winds are still possible directly under these weak tornadoes, and objects in their path such as buildings and cars can be severely damaged. To my knowledge, no tornadoes even close to the severity of the Edmonton and Pine Lake storms in Alberta, which caused complete devastation, have ever been reported in British Columbia.

Several main ingredients are necessary for the formation of a tornado. There must be a hot, humid air mass near the ground with substantial cooling at mid and upper levels of the atmosphere. Also needed is a "cap" that stores the heat and moisture early in the day, followed by a "trigger" such as a cold front arriving to release the energy later in the day. Finally, an element called wind shear, where the wind velocity changes with height, completes the mixture. Changes in wind direction help to increase the overall shear for supercell development and are important for tornado formation.

In summer, when subtropical air masses move up from the south, the criteria for severe storms and possible tornadoes is met several times each season in the Interior. The main reason that tornadoes are rare, though, is that the mountains tend to break up any spinning effects resulting from wind shear.

The Interior Plateau, especially west of Prince George, is a good area for tornado development because the ground is relatively flat. Several farms have suffered damage to buildings and crops over the years.

On October 26, 1994, severe thunderstorm winds caused close to a million dollars in damage around the city of Prince George. Torrential rains and hail caused localized flooding downtown. A small tornado touched down to the west of the city at Clucluz Lake. A damage assessment showed mainly F1 and F2 damage, but there was a small area where the tornado briefly intensified to F3. Fortunately there was little damage to property in the sparsely populated area.

In July 2001, a weak tornado touched down in Cranbrook. Other funnel clouds have been sighted over the years, but this tornado is the only documented one in the Southern Interior. There were several close calls with trees coming down near people, but fortunately no one was injured. Some outbuildings were destroyed and shingles were torn off roofs, but no major damage was reported.

Tornado damage

Top British Columbia Tornadoes

Location	Date	Anecdotal
Cranbrook	Jul. 24, 2001	F1 intensity; winds about 115 km/h; destroyed small out-buildings, tore off shingles, uprooted large trees and threw a fifth-wheel trailer on top of a storage shed
Prince George	Oct. 26, 1994	struck at 4:30 @L ; storm's track was 4 km long and 80–160 m wide; $1 million in damages, including 26 buildings and vehicles struck by flying debris
Cluculz lake and Prince George	Jul. 2, 1991	F1–F2 intensity over short distances and intensified to F3 briefly; struck at 1:30 OL ; storm's track was at least 8 km, mainly over land but 2 sections were over water; hail and heavy rains hampered traffic, 15.4 mm of rain fell in 25 min-utes at the Prince George airport, 2 RVs damaged
Lac La Hache	Apr. 5, 1991	F1 intensity; struck at 11:45 @L ; $16,000 in damages to a motel that was unroofed, uprooted 2 mature Douglas-firs
Tachie on N shore of Stuart Lake	Aug. 13, 1990	F2–F3 tornadic waterspout; struck at 6:45 OL ; $90,000 in damages to buildings
Soda Creek, between Wil-liams Lake and Quesnel	Jul. 5, 1990	F2–F3 intensity; struck about 5 OL ; storm's track was just over 1 km long and 15–530 m wide; many trees uprooted or snapped off, including 1 mature Douglas-fir shattered to the rootball
Prince George	Sept. 3, 1976	F1 intensity; marble-sized hail; damages included 2 mobile homes lifted and moved, a car overturned and downed trees
Castlegar, Blue-berry Creek	Jul. 24, 1974	F1 intensity; struck around 7 OL ; lightning started spot fires; $15,000 in damages, including downed trees, power lines and phone lines, a hangar destroyed and planes damaged—1 blown almost .5 km and lifted over a 1.8-m fence, and another overturned
Duncan, Keating	Apr. 23, 1973	F0 intensity; struck at 2 OL ; hit part of a mobile home park, ripping a carport loose and dropping it on a car
Ucluelet	Mar. 7, 1966	winds reached about 322 km/h; knocked out power, ripped off roofs, sunk boats, smashed windows, snapped wharf pil-ings, picked up a 2-car garage and dumped it into the water, airborne gravel smashed windshields ; 29 cm of snow fell at Marlon Creek and snowdrifts topped 1.8 m
Castlegar, Blue-berry Creek	Jul. 27, 1965	F1 intensity; damaged 4 planes, including a water bomber blown into a hangar and a Cessna lifted up and smashed into a ditch
Prince George	Jun. 9, 1952	the worst tornado in 40 years; damaged houses, ripped off a 23-m section of roofing and tossed it several hundred metres
Spring Lake, 20 km SE of 100 Mile House	Jun. 12, 1926	F1 intensity; storm's track was 0.5 km wide and 80 km long; trees and power lines downed, roofs blown off, cattle killed and crops, fencing and an irrigation ditch destroyed; heavy rain caused flooding, and lightning started many spot fires

Fig. 8-5

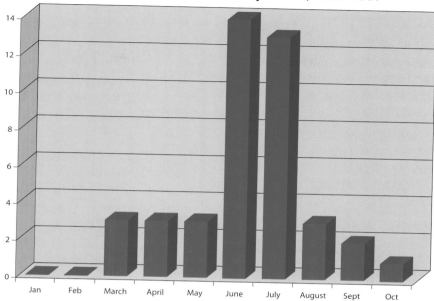

Total Number of BC Tornadoes by Month, 1950–1997

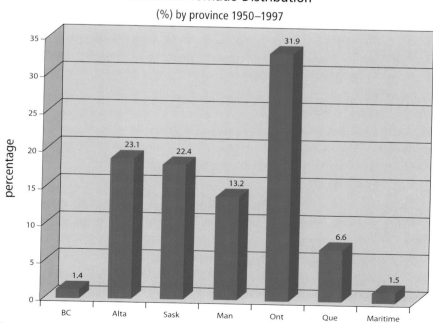

Canadian Tornado Distribution
(%) by province 1950–1997

Fig. 8-6

147

Hailstorms

Hailstorms are almost always brief events in BC, often lasting only a few minutes. However, most years they cause millions of dollars in damage to property and crops. Typical hailstones are about the size of peas, but under the proper conditions they can be much larger.

Hail ranks as one of the most costly natural hazards in Canada. An early September 1991 storm in Calgary resulted in insured losses of $450 million. The damage from hail in BC hasn't come anywhere close to that figure, but there have been some significant storms.

Back on August 4, 1994, one of BC's worst hailstorms pummelled the Salmon Arm area. The early morning was warm and humid with little temperature relief from the previous day's heat. Afternoon thundershowers were in the forecast, but there were no signs that severe storms were brewing.

It stayed mostly sunny until noon, but then clouds developed rapidly, and by late afternoon the sky had turned black. Shortly after, golf-ball–sized hail hit parts of the city. The Hilltop Toyota dealership sustained the worst damage; all the cars in the lot suffered considerable denting. They were sold later with deep discounts. Damage to orchards and property was also very high. The insurance claims for the storm topped $11 million, which is in the top 20 for any hailstorm in Canada.

Orchards in the Southern Interior are very susceptible to damage from hailstones, even from the pea-sized variety. Because the quality standards are set so high for fruit, even very small bruises are not allowed on fruit destined for the fresh market. Late into the growing season, larger fruit such as peaches and apples make easy targets for hailstones, and the whole crop can be ruined in a matter of minutes. As a parttime orchardist in the Central Okanagan back in the '90s, I am well aware of the frustration of inspecting each apple, trying to salvage a few good ones. The other option is to divert the whole crop to juice, where the earnings are one-tenth or less of the money that could have been earned on the fresh market.

There is little farmers can do to protect their crops, especially when they have large acreages. There have been some attempts, though, on a smaller scale to limit losses. Some orchardists in Creston, in the southeast corner of the province, have used netting over their trees to stop the hailstones from hitting fruit. A careful analysis must always be done, though, before these projects are undertaken because the cost of materials may outweigh the value of the crops.

July 1997 Storm

I was working the forecast desk at the Mountain Weather Centre in Kelowna on July 21. The early morning shift change briefing was ominous—a week of hot, dry weather under a strong ridge of high pressure was about to come to a dramatic end. Satellite pictures were showing a band of cloud associated with a sharp cold front lying along the BC south coast. Projections displayed the front on a path to bring it through the Okanagan by mid-afternoon. The overnight convective assessment, which is a process that examines all weather factors affecting storm creation, indicated there

was a very good chance of severe weather developing along the front.

I must admit my attention was divided. I was focused on the job at hand, to consult with my co-workers to make decisions on the timing and extent of weather watches and warnings as the front moved across the Southern Interior. However, I also had a nagging feeling of dread. Along with the prediction of extensive lightning, heavy downpours and strong winds, there was also a high probability for hail. Memories of an August 1994 hailstorm that decimated our apple crop kept running through my mind. All farmers know the feeling of helplessness when there is nothing you can do but wait to see if months of work will be for nothing. There is much hand-wringing, and you chastise yourself for not taking out more crop insurance.

Near noon, the temperature at the Kelowna airport was close to 30° C, and the air was heavy with moisture. By this time, the severe weather watches had been issued, and attention shifted to weather observations, radar and lightning displays. The phone line was ready for calls from volunteer storm watchers.

As soon as there was an indication of severe weather, the watch would be upgraded to a warning. At 2 OL, the western sky turned black, and the first reports of severe weather started to arrive. Between 3 and 4 OL, the front brought torrential rains, strong winds, almost non-stop lightning, much cooler temperatures and hail to most of the Southern Interior. The stones weren't that big, but they were so numerous that about 40 percent of the fresh fruit crop throughout the Valley was deemed unfit for sale. The Central Okanagan was particularly hard hit, with up to three-quarters of the apple crop destroyed. Wind gusts of over 100 kilometre per hour contributed to the devastation.

Damage wasn't limited to orchards. Waves on Okanagan Lake capsized boats, and several car accidents were attributed to the storm. Trees were toppled, bringing down lines and resulting in extensive power outages. All tolled, insurable losses came to $100 million, making it the costliest storm in Southern Interior history. I was one of many who filed an insurance claim, only receiving a fraction back on what the crop was worth.

149

Three Types of Weather Alerts

Special Weather Statements
- issued for unusual weather events that could inconvenience or alarm the general public and that are not described in the weather forecast
- may be issued for events that are occurring outside of but may make their way into the forecast region
- updated as needed

Watches
- issued when conditions are right for a potentially disruptive storm, but when the track and strength of the storm are unknown.
- in summer, may be issued up to six hours before the event
- in winter, issued at least 12 to 24 hours before the event

Warnings
- issued when severe weather is occurring or will occur
- most are issued 6 to 24 hours in advance, except thunderstorm warnings, which may be issued less than one hour before the event
- are updated at least every six to eight hours or as needed

Fig. 8-7

Winter Snowstorms

BC mountains are famous for heavy snowstorms. West-facing slopes in alpine areas commonly receive 50 centimetres of snow or more in a single storm. Higher elevations can experience snow in any month of the year, but heaviest falls from storms are usually in winter through early spring.

The greatest one-day snowfall in all of Canada was at Thatsa Lake (80 kilometres southeast of Kitimat), where 145 centimetres fell on February 11, 1999. Stewart, at the head of the Portland Canal, is another well-known heavy snow location. At the nearby Granduc mine, in a series of storms in the second week of February 1965, 480 centimetres of snow fell. The weight of the heavy, wet snow poised in the Leduc Glacier above the mine site finally let loose at around 10 @L on February 18. The resulting avalanche wiped out almost all of Portal camp. Unfortunately, the men had no warning, and 26 lost their lives in the disaster.

Here is an interesting trivia question—which Canadian capital city west of the Maritimes has the record for the heaviest snowfall in 24 hours? Toronto? Winnipeg? Edmonton? I assume that Victoria was not a place you considered. From December 28 to 29, 1996, an incredible 65 centimetres were measured at the "City of Gardens," while 91 centimetres fell at Colwood, a short distance west of the city. There had already been a couple of significant snowfalls on December 22 and 27, but the storm on the 29th dwarfed everything that had taken place during the previous week.

A cold arctic air mass had been covering the entire province since December 21. By coastal standards, the depth of the cold air over southern Vancouver Island was substantial. When the upper airflow finally shifted to the west, a powerful Pacific frontal system provided the moisture necessary for a heavy snow event. Normally during the transition to a warmer air mass, snow falls for a short time in Victoria before changing to rain. However, with temperatures hovering around -5° C at the start of the storm, heavy snow fell for a much longer time.

Many roofs and greenhouses caved in under the weight of the snow and the rain that saturated the snow a few days later. At Victoria airport, two Viking Air hangars collapsed, resulting in a loss of four aircraft. About 150,000 homes were without power, and many motorists were stranded when the few snowploughs were unable to clear roads. There was over $200 million in total losses.

The storm's impact was not limited to southern Vancouver Island. Vancouver recorded 40 centimetres of snow, and blizzard conditions closed the Trans-Canada highway east of Abbotsford for two days. Numerous avalanches forced authorities to close the Fraser Canyon and other highways to the Interior for many days after the snowstorm.

October 2006 Snowstorm

The official start to winter was still almost two months away when another intense storm hit BC on October 27 and 28, 2006. Moisture-laden Pacific air collided with an unseasonably cold arctic air mass. The combination of warm, moist air aloft overrunning cold air at the surface resulted in an unprecedented early snowfall of between 50 and 150 centimetres across the Central Interior. The nearly stationary maritime front also deposited 300 millimetres of rain on the north coast of BC over several days.

The storm was disruptive to many people, and many were stranded by the snow and cold. Forty forestry workers in Tumbler Ridge and across the border into Alberta spent a few very chilly nights outdoors as low cloud ceilings prevented helicopters from retrieving them. Dozens of hikers and hunters were caught unprepared for the storm, and it is only because of the diligence of Search and Rescue crews that no lives were lost. At least 15,000 people were without electricity in a wide area stretching from Smithers to Fort St. James. Numerous schools were closed, and highway crews worked continuously to keep roads open.

October 2006 Snowstorm Values

Location	Observation
Tumbler Ridge	150 cm of snow Friday evening to Saturday evening
Ootsa Lake	95 cm of snow Friday evening to Saturday evening
Burns Lake	80 cm of snow Friday evening to Saturday evening
Smithers	60 cm of snow Friday evening to Saturday evening
McGregor	150 cm in 30 hours

Fig. 8-8

Blizzards

Blizzards are relatively rare in British Columbia, but when they occur, they can be one of the most dramatic winter storms. The criteria for blizzard conditions in BC include visibility of less than 1 kilometre in snow or blowing snow, winds of 40 kilometres per hour or greater and temperatures less than -10° C (-5° C on the coast). These conditions must persist for at least six hours. The amount of snowfall is not included in the definition—blowing snow is the issue. But even if severe winter storms do not meet the criteria of the official definition, they are not to be taken lightly. Blizzards not only cause hardship, they can also bring loss of life. Environment Canada's website warns, "Winter storms and excessive cold claim more than 100 lives every year in Canada, more than the combined toll from hurricanes, tornadoes, flood, extreme heat and lightning."

Southern BC was hit by what the media called "The Blizzard of '96" a few days after Christmas of that year. The storm produced heavy snow, low visibility and cold temperatures, but it wasn't a true blizzard because it lacked one main part of the blizzard definition—wind. Only the eastern Fraser Valley and the Fraser Canyon experienced true blizzard conditions. The storm did, however, disrupt lives and cause hardship to millions of people for many days after the event.

Freezing Rain Storms

Freezing rain in BC is most often found in a fairly narrow band associated with warm, moist Pacific air overriding a cold arctic air mass. The most common occurrences are in the Peace River country and the eastern Fraser Valley. Chilliwack and Agassiz are prime locations for freezing rain. With a retreating arctic air mass, it will typically be snowing in Hope and raining in Abbotsford, but freezing rain continues to fall between the two communities for hours or even days at a time. An ice storm in January 1972 caused considerable damage to power lines, and similar to the 1986 storm in southern Quebec, many steel transmission towers came down. Hydro crews chipping ice to climb poles estimated that the ice coating was 5 to 7 centimetres thick! Many branches snapped like sticks, and even whole trees tumbled with the terrific weight upon them. A "silver thaw," as some call it, covered the entire countryside with a silver-like glaze. Fortunately storms of this intensity don't happen that often, but almost every year, there are brief periods of freezing rain somewhere in the province.

Vancouver on December 21, 2008

Chapter 9:
Observing the Weather

Old weather saying:
Rain before seven,
fine by eleven.

What will the weather be like tomorrow? Next Friday? This time next year? These are questions that we ask. Looking out the window gives us our first impression as to what the weather will do for the next few hours, but if we are interested in what is going to happen tonight or tomorrow, looking out the window is inadequate. In the 19th century, when people first undertook an intensive, organized study of the atmosphere, they realized there could be very different conditions at two locations, even when the locations were not far apart. This realization led to the understanding that weather is organized into patterns and that these weather patterns move. So it was concluded that observations should be taken at many locations at about the same time and then gathered together at a central location where the patterns could be analyzed. Then a forecast of future weather conditions could be prepared based on an understanding of the rules or laws that determine how the atmosphere works.

The nature of weather and how it behaves on different scales is based on the principles of fluid dynamics. The atmosphere is a fluid that is in turbulent flow on a rotating sphere. Mathematician Lewis Fry Richardson may have summarized it best when, early in the 20th century, he described turbulent fluid flow: "Big whirls have little whirls that feed on their velocity, and little whirls have lesser whirls and so on to viscosity." This ditty recognizes some basic realities of the atmosphere: that it is a turbulent fluid that behaves on many different scales of motion, and that there is a link between these scales of motion.

> *Because measurements of weather parameters are taken from all over the world, the information is recorded in a special code that can be understood by the people of all countries, regardless of the language they speak.*

So, the essence of weather forecasting is to measure the smallest scales reasonable and use the rules of atmospheric motion that define the link between these scales to give a picture of the state of the fluid at some time in the future. Weather patterns have a space scale of a few hundred kilometres, and they can change over a few hours. To resolve these weather patterns requires a network of stations spaced about 100 kilometres apart that take observations every hour. This space and time scale is common among most meteorological observing programs that support weather forecasting.

Observation Systems

Assembling observations and preparing forecasts for even a few locations requires a large-scale, coordinated approach. Consequently, the observing and forecasting systems are largely the responsibility of national organizations. The data-observing approach must be standardized to make the international exchange of data run smoothly. Observational standards follow an international protocol established and maintained by the World Meteorological Organization, which operates under the United Nations. The data is communicated to one or more locations, where it can be analyzed so forecasts can be prepared.

Weather forecasting requires data from the full depth of the atmosphere. In Canada, the surface weather observation network consists of several hundred stations across the country that repeat measurements, generally on an hourly basis. In addition, about 30 stations measure meteorological variables above the surface to heights of about 30 kilometres. At these aerological stations, surface weather data is also measured, and many other specialized instruments are used for specialized observations that support atmospheric research programs. These surface and upper air *in-situ* measurements are supplemented with data from remote sensing systems such as weather radar and weather satellites. The data is sent to local and national weather offices, and much of it is made available to the public through the internet and the media.

Gathering marine weather data

Surface Weather Observation

Formal weather observations were recorded at locations across British North America through the 18th and early 19th centuries. The first official observation program began on Christmas Day in 1839. According to a plaque on the University of Toronto campus, the British Army began regular meteorological and magnetic observations on the site, then called Her Majesty's Magnetical and Meteorological Observatory, in 1840. The first weather observation network was established in Great Britain in the 1850s. Canada's first network was established in 1872. It covered southern Ontario, Quebec and the Maritime provinces, and data was communicated using the newly installed telegraph systems. Esquimalt Naval Base, near Victoria, has the earliest recorded weather data in BC, dating back to 1872. New Westminster and Spences Bridge began observations a few years later. The old gold rush town of Barkerville has the longest continuous

set of records, starting in 1888. The 120 years of weather data is very useful in the study of climate change. At present, the Meteorological Service of Canada operates the national weather observation program that supports the preparation of weather forecasts.

The measurement of atmospheric variables has a long history, and electronic technology and computerization have influenced weather observing methodologies. Meteorology has evolved into several areas of specialization, and the monitoring requirements have evolved as well. Data requirements for weather forecasting are somewhat different than those needed for atmospheric research or for specialized applications in air quality. Weather forecast requirements are based on what are known as standardized observing protocols. Issues including instrument location, representativeness and accuracy of measurement, and observation averaging times must be considered if data is to be used for the forecast. Predicting the atmosphere is difficult enough without the need to take into account idiosyncrasies for each weather measurement.

At weather observing stations, the instruments are located where they provide a measurement that is representative of the atmosphere. For example, they are not sheltered by a tree or building, a temperature reading is not artificially elevated by direct sunlight or the heat rising off a paved surface, snow measurement is not subject to a snowdrift depending on wind direction, and a precipitation measurement is not elevated by rain dripping off a roof or swirling around a tree.

Most observations are from unstaffed sites where data is measured automatically. Many sites are located at airports, where trained observers provide additional data that is useful for the aviation industry.

Many instruments are housed in a ventilated cabinet known as a Stevenson screen. Named after its inventor, Thomas Stevenson (father of the famous author Robert Louis Stevenson), the cabinet ensures that sunlight does not fall directly on temperature gauges. The cabinet is painted white to ensure that the sun's rays do not cause it to heat.

Each Stevenson screen contains different types of thermometers.

Precipitation

The history of precipitation measurement goes back to the 4th century BC in India. The principle used then is the same one used now—measure the amount of precipitation collected in an open-topped container. Refinements have been focused on reducing the amount of precipitation lost from splash-out of rain drops, reducing the effects of wind on collection efficiency of the opening, and reducing evaporative loss between precipitation events. The manual rain gauge used in Canada has a design that has been largely unchanged since 1871. The rate of rainfall is measured by a tipping bucket rain gauge. Collected rain passes into a small container that empties each time its capacity is reached. The rate at which the container empties indicates the rainfall rate.

A tipping bucket rain gauge (above); the tipping bucket (below)

Snow measurement has been a challenge. The snow catch rate of a collection container is greatly influenced by wind, so most devices have a shield to reduce wind speed around the opening. The shape of the opening also influences how efficiently a container collects snow. The Canadian Nipher-shielded Gauge has a bell-shaped opening to minimize loss. Where there is a human observer, the depth of snow accumulated on the ground is determined by a simple depth measurement at a number of sites in a so-called snow course. Measurements are taken at several locations so differences in accumulation that result from wind drifting can be averaged.

157

Many innovative techniques have been used to measure snowfall automatically. Snow pillows laying on the ground measure the weight of the snowpack that accumulates during winter. The storage gauge collects snow in a large container filled with glycol, which melts the collected snow, and measurements of the weight of the container are taken periodically. The depth of snow can also be measured at one point on the ground by a downward-pointing microwave sensor.

Automated detection that identifies the type of precipitation is difficult. One approach measures the fall rate of the precipitation by means of a small Doppler radar and links the fall rate to the type of hydrometeor. This approach works well enough to differentiate between snow and rain, but discriminating between drizzle and snow is not easy because they have essentially the same terminal velocity. The detection of freezing precipitation is also a challenge. One approach uses a vibrating probe and calibrates the rate of vibration to the rate at which precipitation adheres.

Temperature

Around 1592, Galileo invented the first thermometer, which used the expansion of air to measure temperature. However, it was prone to many errors, and the apparatus was large and cumbersome. Scientists soon realized that thermometers made of sealed tubes containing mercury or alcohol were much more reliable and practical.

Gabriel Daniel Fahrenheit developed a temperature scale in 1724 that had three fixed reference points. Zero was the value given to the coldest temperature that he could create by mixing ice, water and ammonium chloride. The upper end of the scale was the temperature of the human body, which Fahrenheit arbitrarily called 96 degrees. The third fixed point was the temperature of a water/ice mixture that, on his scale, had a value of 32 degrees. The Celsius temperature scale, invented by the Swedish astronomer Anders Celsius in 1742, is the one most frequently used worldwide. On this scale, the temperature of the water/ice

Microwave snow-depth sensor (above);
Nipher-shielded Gauge (left)

Maximum and minimum thermometers in a Stevenson screen

mixture was set as zero, and the value of 100 degrees was assigned to the temperature at which pure water boils.

Various types of thermometers are used at surface weather stations. The minimum thermometer contains a tiny metal dumbbell (index) that floats in liquid, usually alcohol. The index is forced toward the bulb by the retreating surface of the liquid as the temperature falls. When the temperature rises, the dumbbell remains in place and registers the minimum temperature. The maximum thermometer has an index that is pushed upward in the barrel of the thermometer and registers the maximum value reached in the observation period. Regular, maximum and minimum thermometers are all manually read.

Temperature sensors have been developed for use in automated systems or where space limitations or working environments are unsuitable for liquid-filled thermometers. The most common are instruments known as thermistors (the name is a combination of "thermal" and "resistor"), which are made of mixed metal oxide semiconductors. In a semiconductor, the resistance to the flow of electrical current changes as the temperature of the device changes. In some thermistors, the resistance increases with increasing temperature, and in some, the resistance decreases with increasing temperature. The semiconductor devices can be physically small and rugged in design, which makes them suitable for use with electronic circuitry.

A sling psychrometer to measure humidity

Humidity

Humidity measurements are important in meteorology because knowledge of the moisture content of air is necessary to determine the formation of cloud. In 1660, Francesco Eschinardi demonstrated that the evaporation of water causes a thermometer to cool. The psychrometer is a device that uses this principle to determine the amount of moisture in the air based on the difference in temperature between two thermometers. The dry bulb thermometer measures air temperature normally, whereas the so-called wet bulb thermometer has its bulb enclosed by a water-saturated wick. A fan draws air past both thermometers at a known rate, and the temperature difference is then used to determine the humidity based on previously calibrated values.

The psychrometer is not suitable for automated measurements of atmospheric humidity because the water reservoir would need to be maintained, and the behaviour of the device changes at temperatures below freezing. An instrument called the dewcel is used at Canadian weather stations. This device has a fibre-glass sleeve that is saturated with a lithium chloride solution. The lithium chloride absorbs moisture from the air, and a heater is used to raise the temperature of the sleeve, causing the moisture to evaporate. A thermistor measures the temperature difference between the heated and unheated sleeve. This temperature difference is proportional to the atmospheric moisture.

Many other devices have been developed to measure humidity. They use different principles to measure moisture, such as the change in length of hair as it absorbs water vapour, the change in electrical properties of materials as they absorb water vapour, or the temperature at which a cooled reflective surface becomes fogged because of water vapour.

> Canada's Alert weather station is the world's most northerly. It is located on Ellesmere Island, 830 kilometres from the North Pole.

Pressure

Atmospheric pressure is a result of the weight of overlying air. Surface pressure has traditionally been measured by the mercury barometer (meaning "weight meter"), which Evangelista Torricelli invented in 1644. The modern instrument consists of a mercury-filled tube that is inverted and immersed in a mercury-filled cistern. Atmospheric pressure pushes down on the top of the cistern and supports the column of mercury in the tube. The height of mercury that is supported essentially balances the weight of the atmosphere above the cistern. At sea level, this amounts to the weight of about 76 centimetres of mercury. This type of atmospheric measurement is the most consistent and reliable available. However, the mercury barometer is not easily transportable and does not lend itself to automation, and there are health concerns associated with using mercury. As a result, the aneroid barometer, invented by Lucien Vidie in 1843, is widely used.

Mercury barometer at an airport (above); aneroid barometer with graph to record pressure changes (below)

The aneroid barometer consists of a partially evacuated and sealed metal bellows canister that expands or contracts in response to atmospheric pressure changes. Mechanical gears or electronic techniques determine the amount of this deflection.

> *Flying insects stay closer to the ground when the air pressure drops, so when you see swallows flying low to the ground, it could be an indication that rain is on the way.*

Aneroid systems are used in portable barometers, altimeters and pressure recording devices (barographs). Other pressure measurement techniques using quartz and silicon-based cells are used in specialized applications. To measure atmospheric pressure, the barometer must be located in a building where the temperature is controlled for the stability of the instrument. To ensure the pressure represents outside conditions, the building must not be sealed, and it must be designed to minimize the effects of spurious pressure fluctuations caused by wind pressure, vehicles moving nearby and even the opening and closing of doors.

Wind

Wind may be the most influential weather phenomenon throughout the ages. The Tower of the Winds in Athens was built by Andronicus of Cyrrhus around 50 BC to recognize the eight quadrants from which the wind blows. A weather vane was mounted on top of the tower. Since that time, various types of wind vanes have been developed to measure wind direction, and several different methods have been used to measure the magnitude of the wind. One early approach attempted to measure wind force directly using a suspended flat plate that was deflected by the wind. Such devices date back to about 1450. Subsequent variations on this theme consisted of a flat plate, either square or circular, which was kept facing the direction of the wind by a vane. One such device invented by A.F. Osler was installed at the observatory in Greenwich, England, in the mid-19th century.

Osler's wind measurement device

A more useful measurement concept is wind speed—the distance that air moves over a short period of time. Wind speed is often measured with the cup anemometer, which operates on the principle that cups mounted on arms rotate around a vertical axis as the air moves past. The number of rotations in a fixed interval of time is proportional to the wind speed. John Thomas Romney Robinson invented the first such instrument in 1846, and Canadian John Patterson invented the common three-cup anemometer in 1926. The anemometer and wind vane are generally mounted on a tower that meets an international standard for

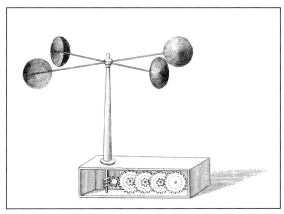

The Robinson cup anemometer

height and siting. Measurements are taken 10 metres above ground, and the tower is located where the wind flow will not be influenced by any trees or buildings.

A standard wind vane with the Patterson three-cup anemometer

A propeller-and-vane anemometer

measures the speed at which an aircraft moves through the air, but it is also used in ground-based instruments to measure rapid fluctuations in wind speed. The sonic anemometer calculates wind speed by measuring the time taken by two sound pulses to move in opposite directions across a known distance. The difference in time that the two pulses travel is proportional to the speed of the air. This instrument has the advantage of no moving parts and can measure rapid changes in wind speed.

> **It was one of those March days when the sun shines hot and the wind blows cold: when it is summer in the light, and winter in the shade.**
> —Charles Dickens,
> *Great Expectations*

Other wind measurement approaches have been developed for special applications. At many automated weather stations, the wind speed and direction devices are combined into one instrument. This device has a propeller mounted at the front of a wind vane. Non-mechanical devices are used to measure rapid fluctuations in wind speed and direction. Another approach uses instruments known as hot-wire and hot-film anemometers to measure the cooling effect of air as it moves across an electrically heated surface. A stronger wind removes more heat and, therefore, more electrical energy is required to heat the surface. Yet another approach measures the pressure drop that occurs when wind flows across a small opening. The pressure tube approach was first developed in the late 18th century, and a practical version developed by William Dines in 1892 is still used today. This approach

Visibility

This parameter is important for such things as aviation and marine transportation. Visibility is a function of atmospheric conditions, and it also depends on the capability of the human eye, making it one of the more subjective atmospheric measures. A number of instrumentation methodologies have been developed to bring a degree of objectivity to this parameter. Most instruments measure the drop in intensity of a light source when observed over a fixed distance in the atmosphere. These instruments, called transmissometers, are used at airports to measure the runway visual range, which is critical for aircraft movement around airports.

Observations at Sea

A major international cooperative program is in place to gather weather information from over the oceans. Observations taken on vessels use much of the same types of instrumentation used at land stations. However, because the ships are moving, the wind measurements must take into account the vessel's motion. In addition, the ideal siting criteria used at land stations cannot be met in the cramped conditions onboard most vessels. The key measurements of temperature, pressure and wind are still useful in the weather forecasting programs.

A visibility sensor

Marine buoys are the oceanic version of the unstaffed automatic land stations. There are two types of buoys—moored and drifting. The moored buoys have a long tether fixed to the ocean floor. The observation program usually includes pressure, temperature and wind as well as marine measurements (such as sea surface temperature) and sea state information (such as wave period and height). Occasionally, the buoys also collect ocean current information. Drifting buoys are generally limited to temperature and pressure measurements, but some gauge wind speed as well. Data is usually communicated through the meteorological satellites to the world communication network. Although the location of the moored buoys is known, the drifting buoys must be tracked using global positioning technology. The motion of the buoys provides a measure of the ocean current.

Measurement Aloft—the Aerological Monitoring Program

Weather forecasting programs need regular measurements of the upper atmosphere. These aerological measurements, as they are known, are taken year-round every 12 hours, at noon and midnight Coordinated Universal Time (that's 4:00 and 16:00 Pacific Standard Time in BC). The spacing of the upper-air observing network is much wider than the surface observing network, and stations are about 600 kilometres apart. In BC, there are four stations in the national network; they are located at Port Hardy, Kelowna, Prince George and Fort Nelson. Measurements are taken using a lightweight instrument package that is lifted using a helium-filled balloon. The instrument

Releasing a radiosonde

earth. These instrument packages are not normally recovered. The balloon-borne observations are taken in all weather conditions. It is particularly challenging to release the balloon when the surface winds are strong and freezing rain or blowing snow is occurring.

Once a week, ozone measurements are taken at 10 Canadian aerological sites. This monitoring program focuses on ozone in the stratosphere and the depletion of this layer, which protects life forms from the harmful effects of ultraviolet radiation. At present, Kelowna is the only location in BC where the ozonesonde is released. Ground-based instruments also measure ozone with remote sensing techniques.

Although the aerological network is confined primarily to land stations, a few ships release radiosondes at fixed locations on the oceans. These programs are expensive to operate, and they are now being replaced by programs operated from ocean-going transport vessels, which release radiosondes while en route.

Remote Sensing

Weather observations using instruments that are in direct contact with the atmosphere are known as *in-situ* measurements because they have sensors in direct contact with the atmosphere. Remote sensing techniques probe the atmosphere from a distance.

Various techniques are used for remote sensing, including sound, the reflection of electromagnetic waves (from sunlight, microwave and laser instruments) and the sensing of heat radiated by a substance.

package, known as a radiosonde, includes sensors that measure pressure, temperature and moisture. It has a small radio transmitter on board that sends the data and also allows the motion of the instrument to be tracked at the receiving ground station. The horizontal distance moved by the instrument package during its ascent, determined by using GPS navigation, is used to calculate the wind speed and direction. The balloon bursts at a height of about 30 kilometres above ground level, and a small parachute slows the instrument package in its fall back to

Weather Radar

When radar (an acronym for radio detection and ranging) was first used to track the motion of aircraft, the returns from clouds and precipitation were considered detractive noise. However, it quickly became apparent that radar could be used to provide new information about the atmosphere. Weather radars have a transmitting dish that rotates and sends pulses of radio waves in the microwave frequencies. The dish focuses the pulses along a narrow beam that reflects off cloud and water droplets. When part of the microwave pulse bounces back to the antenna, the radar's electronics time the return and measure its intensity. One-half of the total time for the pulse to travel to and return from the target multiplied by the speed of light determines the target's distance from the radar dish.

A radar site at Silver Star in the Okanagan (above)

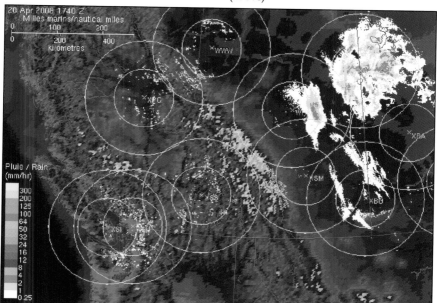

Radar image for western Canada (the April 2008 snowstorm)

Weather Radar Interpretation Errors (adapted from Environment Canada website)

1. Blocking Beam

Uneven terrain such as hills and mountains can block a radar beam, leaving gaps in the pattern—this is very common in the mountains.

2. Beam Attenuation

Storms closest to the radar site reflect or absorb most of the radar energy, so little energy is left to detect storms that are farther away.

3. Overshooting Beam

When clouds are close to the ground, as they are in lake effect snow squalls, the radar beam may overshoot them so that even clouds with intense precipitation will only have a weak echo.

4. Virga

The radar beam detects precipitation that is occurring aloft, but the precipitation never reaches the ground because it is absorbed by low level dry air conditions.

5. Anomalous Propagation

During an inversion in the low atmosphere, when a layer of warm air overlies a layer of cooler air, the radar beam can't pass between the two layers and is reflected back to the ground, sending a false strong signal back to the radar site—an event that is most common in the morning.

6. Ground Clutter

The radar beam echoes off objects on the ground, such as tall buildings, trees or hills.

The strength of the reflected signal is proportional to the size of the reflecting object. Microwave radiation reflects off cloud and rain drops (and even swarms of insects). Based on the physical properties of raindrops, the intensity of the returned signal is proportional to the size distribution of the reflecting droplets. Using rain gauge data, the rainfall rates are correlated with the assumed droplet size. In this way, the rainfall rates in the sensed clouds can be estimated. Canadian weather radars transmit in the 5-centimetre wavelength and have an effective viewing distance out to about 250 kilometres. The radar display is a circular pattern centred at the radar site. Usually a sequence of images is displayed, which then shows the motion of the echo. The Canadian weather radar network is operated by Environment Canada and includes four sites in British Columbia.

Most radars also make use of the Doppler principle to measure the shift in frequency of the returned echo. Changes in the frequency of the returned signal are proportional to the speed that the reflecting object is moving toward or away from the radar site. This Doppler principle is used to measure the flow patterns in the atmosphere. It is useful for examining the wind patterns in intense thunderstorms, and the Doppler radar can even be used to determine if the cloud echo is rotating. A rotating cloud echo usually indicates a severe thunderstorm, which may generate a tornado even though tornadoes themselves are below the resolution of the radar beam. All the radar sites in BC have Doppler-processing capability.

The Doppler Effect

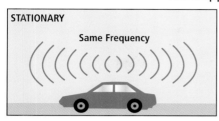

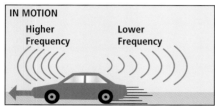

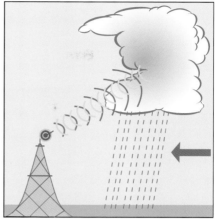

Fig. 9-1 The frequency of sound waves is shifted because of the motion of the source. The frequency of microwave radar echoes is shifted because of motion of cloud and rain.

Weather Satellites

The weather satellite is a relatively recent invention, but it has become one of the most useful tools for observing large portions of the globe. The TIROS satellite launched by the United States in 1960 was the first, followed by scores of satellites that have since been launched by governments and other agencies.

Most weather satellites use passive sensing systems, which receive two basic types of radiant energy. Visible light produced by the sun is reflected off the earth's surfaces and clouds, back up to the satellite. Visible sensors on meteorological satellites are essentially black and white cameras. Clouds look white to the satellite-based sensor, whereas ground surfaces and water bodies appear grey or black. A second type of sensor detects infrared or heat energy. The intensity of the infrared energy is related to the temperature of the emitting surface, a phenomenon known as Boltzmann's law.

The satellite infrared sensors are sensitive to the temperatures of the earth's land and water surfaces as well as cloud tops—a temperature range extending from -70° to 60° C. Because the earth and atmosphere emit heat day and night, infrared images are always available, unlike visible images, which are only available during daylight.

There are two types of satellite systems now generally in use. Geostationary satellites are placed in circular orbits over the equator at heights of 35,800 kilometres. The satellite moves in the same direction as the earth rotates and takes one full day to orbit the earth. It basically appears stationary above the earth's surface. The second type of satellite is in a much lower orbit at about 800 kilometres above the surface. The orbit passes near the poles and is at a slight angle of inclination relative to the equator. The orbital geometry is such that the satellite orbital plane moves in time with the sun.

169

Weather Satellites

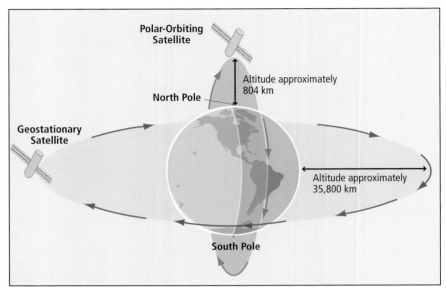

Fig. 9-2 Polar-orbiting satellites are close to earth and provide high resolution images. Geostationary satellites are in high orbits and provide low resolution images.

These polar-orbiting satellites take a continuous swath of images along their orbit and, as a consequence, the images appear as one long, continuous image. Because geostationary satellites are about 45 times higher than polar-orbiting satellites, the resolution is lower, and images from the equator-oriented satellites become greatly distorted toward the north and south poles. The polar-orbiting satellites move around the earth in about 100 minutes, and most places on the earth's surface are scanned twice daily, once in daylight and once in darkness. Because they pass frequently over the poles,

A polar-orbiting satellite image

A geostationary satellite image

these satellites provide more useful data at those latitudes than do the geostationary satellites. The infrared sensor on the geostationary weather satellites can distinguish areas that are 4 kilometres in width, while the images in the visible light spectrum can resolve objects as small as 1 kilometre.

Satellite observations have an obvious advantage over areas of the earth for which there is little conventional weather data available. Tracking weather systems as they move across these areas has proven to be immeasurably valuable. Satellites are, in fact, used to produce "synthetic soundings" of the atmosphere over ocean areas where there are few aerological stations.

Acoustic Sounders

These instruments, called SODARs (for sonic detection and ranging), use a pulse of sound to measure wind in the lower atmosphere. Their principle of operation is similar to that of radar except that they use sound waves rather than microwaves. When a sound pulse meets a discontinuity in the atmosphere, sound is scattered in all directions. The discontinuity can be in the temperature field, such as a strong temperature inversion. There can also be discontinuities in wind, associated with eddies that are present in the turbulent air flowing over the earth's surface. The SODARs' speaker directs a burst of sound, and a receiver then listens for a short period of time for the faint sounds that are scattered. The time from the initiation of the sound burst to the receipt of the backscattered sound provides an estimate of the distance to the discontinuity, and the intensity of the return is a measure of the discontinuity's strength. The sounding systems can also detect slight changes in the frequency of the sound, and this Doppler shift is used to measure wind speed. Finally, the systems allow wind profiles to be obtained over the lower atmosphere to a maximum height of about 1.5 kilometres.

Lightning Detection Networks

Recently, lightning detection systems have been developed that use the electromagnetic wave properties of a lightning discharge to determine the location of the lightning stroke. As is readily apparent when listening to an AM radio during an electrical storm, flashes are accompanied by an immediate static burst on the radio. The strength of the noise (i.e., its amplitude) depends on the proximity of the radio receiver to the discharge. A network of detection stations has been established across North America. The detector consists of an antenna that can determine the direction (or azimuth angle) of a lightning discharge from the antenna. Using the azimuth angles from several of the closest antennae, the discharge's location can be triangulated.

The system in use has a network with detector spacing of 200 to 400 kilometres, which gives it a data collection efficiency of up to 90 percent and a location accuracy of between 8 and 16 kilometres. Lightning strike information is analyzed by dedicated computing systems, and the location and density of discharges are displayed and updated every hour on a surface map. The movement of areas of lightning strikes and changes in the discharge rates can provide valuable information on the speed and rate of intensification of thunderstorms.

Communication of Meteorological Data

Availability of appropriate weather data at a central location for analysis, forecast preparation and dissemination all depend on an accessible and reliable communication system. The invention of the telegraph in 1837 presented the first opportunity for meteorologists to undertake weather forecasting programs that could be useful to the public. Dedicated communications networks are used to send data to weather offices within minutes of the collection and analysis of the observation. The internet has made the communication task much less burdensome and has allowed the public to gain access to observations, analyzed data sets and all types of forecasts.

Data Quality Control

Quality control of the observational data is essential to the weather forecast production system. Weather monitoring instruments are maintained and periodically calibrated to ensure that their operation is stable, and that measurement results are repeatable. At one time, trained weather observers and technical staff overviewed the data, but today computers do most of the work. Observations are stored in data banks, where quality control computer programs check for observational range, consistency with previous observations from the site and consistency with nearby observations.

Chapter 10:
Forecasting the Weather

Government organizations and the private sector both undertake weather forecasting programs. The Government of Canada issues weather forecasts that "are provided for the good of the general population," according to federal policy. Governments worldwide provide a similar service for their populace. Provincial governments are responsible for forecast services that support sectors under their jurisdiction; the BC government operates weather forecast programs to support forest fire suppression as well as stream flow and flood warning services. Prediction services operated by the private sector serve the specialized needs that fall outside the government's role. The main focus of this chapter is the federal forecasting process.

Weather forecasts are prepared and disseminated when they will be the most useful for clients. Warnings of hazardous weather conditions are prepared to enhance public safety and reduce property damage. Forecasts are also prepared to support many sectors of the economy, including transportation (aviation and marine transport), recreation (boating, skiing, avalanche warning) and construction. The public wants to know what weather conditions to expect for the day, and the weather forecast is a regular item on newscasts and in newspapers. In Canada, federal weather forecast centres are located across the country to partition the geographic area into manageable portions. These centres operate similar programs, which include some of the following services:

• preparing public forecasts and bulletins for cities

• issuing warnings for severe weather conditions

• preparing special sector-specific forecasts (transportation, recreation)

• providing information about unusual weather

• providing the latest weather conditions and historical climatic data.

Prediction on the Large Scale—Guidance to the Forecaster

The weather forecast problem is complicated because the atmosphere extends well beyond the local forecasting area. Predictions of the atmosphere at the hemispheric and global scale are prepared at a central prediction centre and are sent to local weather offices. Meteorologists use this guidance material to prepare weather forecasts for several days into the future. In Canada, the central prediction facility is located in Montreal. Although this facility is mandated to support the national weather forecasting program, much of the meteorological information it prepares is also made available to provincial and private forecasting offices.

Numerical Weather Prediction

The atmosphere is a continuous fluid that can be described using a series of mathematical equations that link temperature, pressure, wind, density and moisture. In 1904, the Norwegian meteorologist Vilhem Bjerknes recognized that these equations could be used to predict the future state of the atmosphere. In 1922, the British meteorologist L.F. Richardson first proposed a computational process to predict the weather using a simplified version of the theoretical equations. He proposed that a short-range forecast could be produced by several groups of individuals using arithmetic calculators to do the many necessary computations. One individual would coordinate the exercise, acting much like an orchestral conductor, ensuring the groups were working together to make the calculations faster than the actual weather was changing. Other individuals would take interim forecasts and make them available to users. Richardson envisioned having a warehouse to store the results for future examination and a research group to develop new forecasting techniques. He actually tried to prepare a six-hour forecast of pressure change using a version of this approach. After much laborious calculation, Richardson's prediction was off by a factor of about 100 because he did not appreciate some of the basic computational techniques that must be used in the process. The massive calculation exercise that was required and the resulting lack of success meant that further exploration of the technique was essentially abandoned for over a quarter of a

Global Weather Prediction Grid

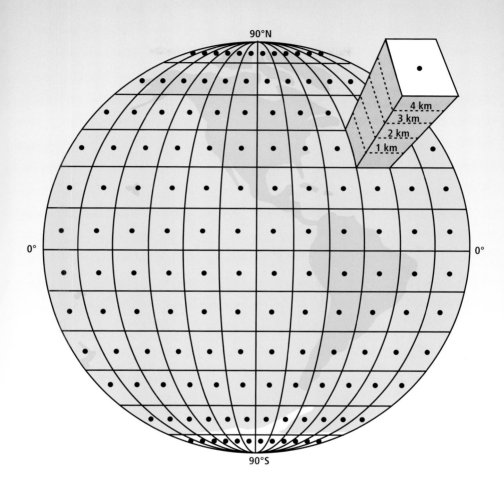

Fig. 10-1 A coarse-mesh grid covering the earth's surface

century. However, Richardson's foresight is now considered quite remarkable.

The full set of equations needed to prepare a weather forecast is too complicated to be solved by simple mathematical techniques, and large computers must be used. The computer-based approach is known as numerical weather prediction

(NWP). In theoretical physics, the NWP approach is what is known as an initial-value problem: if the initial state of the atmosphere and the equations that govern its motion are known, its future state can be predicted. The equations must be in a form that allows observed data to be incorporated, which is accomplished

175

through a process called discretization. In the atmospheric NWP model, a three-dimensional grid of small cells is used to represent the atmosphere. Each grid cell has a horizontal length of a few tens of kilometres and a thickness ranging from tens of metres to a few kilometres. A network of several million grid cells is used in the largest and most sophisticated NWP models, which predict the atmosphere over the entire globe.

Before modern-day weather forecasting, many people took their cues from nature. For example, when they saw ants moving to higher ground, people knew that the air pressure was dropping, so rain could be on the way.

This three-dimensional grid needs to be initialized with data. Using data from observation stations, temperature, pressure, moisture and wind values are assigned to grid points located at each corner of each grid cell. At points above ground level, temperature, moisture, wind and pressure values are assigned using measurements from radiosonde balloon ascents and from instruments onboard aircraft. Synthetic data generated from weather satellites is used over areas largely devoid of observations, such as the oceans and polar regions. The rate of change of each of the atmospheric parameters is calculated using the equations of motion. At each grid point, a new value of the atmospheric parameter is calculated by multiplying the rate of change by a small time step that is a few

minutes long. A short-range forecast of all the parameters is produced, and from this a new rate of change is calculated and another short forecast is computed. This time-stepping process is repeated to generate forecasts out to a few days. The length of each time step is related to the size of the grid cells—the smaller the cells, the smaller the time step.

NWP was only made possible when large numerical computers became available after World War II. The first successful experiments in NWP were completed at Princeton University in the late 1940s. Although it required almost 24 hours to prepare a one-day forecast using a simple model of the atmosphere, NWP's potential was demonstrated. Since the middle of the 20th century, most national weather predicting programs have used the NWP approach. The NWP computing centres use the fastest supercomputers available, and it still takes several hours to generate all the numerical forecasts. The NWP process is usually repeated twice daily, synchronized with the main 12-hour global weather observation cycle, and the guidance material is sent out to weather forecasting centres.

Canada is highly respected in the NWP research field. Since 1963, experts at the Canadian Meteorological Centre in Dorval, Quebec, have been world leaders in the continuous improvement of computer simulations of weather patterns. In 1992, I toured the four-storey facility and was impressed by the dedication of the staff members and their constant quest to find new ways to improve the models.

Numerical Weather Prediction Charts

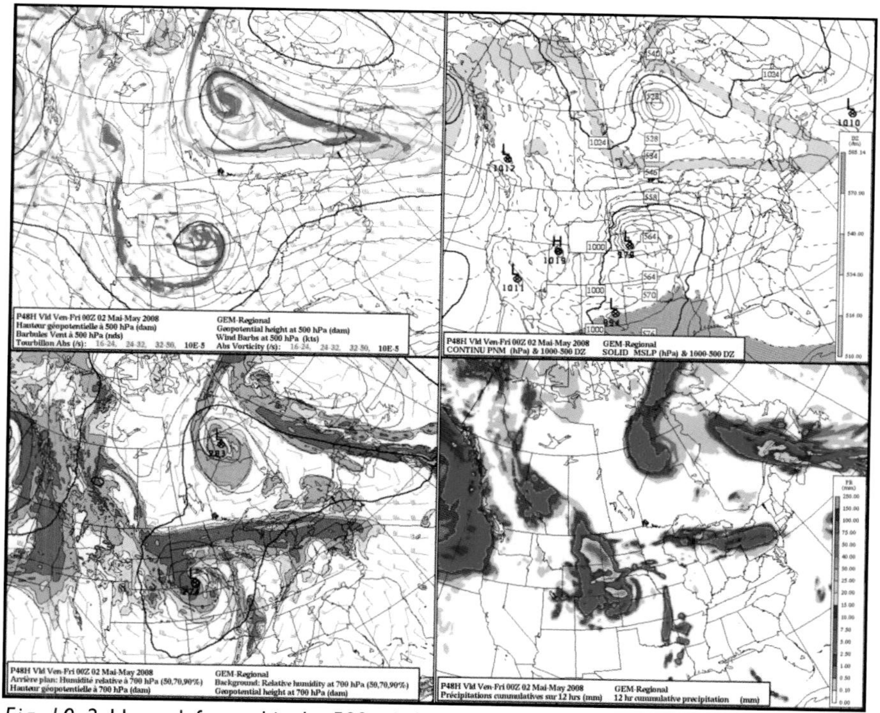

Fig. 10-2 Upper left panel is the 500 mb forecast. Upper right panel is the surface pressure forecast. Lower left is the 700 mb forecast. Lower right is the precipitation forecast. All forecasts are 48 hours.

The Forecast Products

Warning Program

The cornerstone of the weather forecasting service is the warning program, which issues messages alerting the public of potential weather hazards. Watches or warnings are issued depending on the expected severity of the anticipated event. A warning is issued when a weather event is expected to happen in a specific area, and a majority of people in the area may be affected. The rules under which a warning is issued are developed beforehand and are based on public safety considerations as well as the specific needs of weather-sensitive sectors of the economy. Warnings are "short fuse" situations in which the atmospheric state is expected to change quickly. Severe thunderstorm warnings are issued two hours in advance of the event. Watches of severe, large scale events such as blizzards or wind storms are issued 12 hours in advance. The weather watch is a "heads up" that is issued when the forecaster expects that a severe weather event may occur in the near future. Watches are issued up to six hours in advance of a

177

severe summer storm, and up to 24 hours in advance of a winter storm. The forecaster also prepares special weather statements to provide additional explanations of weather conditions that will have a public impact but are not expected to reach the warning or watch criteria. (See Figure 8-7, for more information.)

Aviation

The aviation industry requires detailed forecasts of many atmospheric parameters, and several different products are prepared. The Terminal Aerodrome Forecast (TAF) is used to enhance aircraft landing and takeoff safety at airports. The TAF includes wind speed and direction, precipitation type and intensity, type of visibility obscurations (fog, precipitation, smoke and dust) and height of cloud above ground level at the airport. The forecasts are issued every six hours, out to 12 or 18 hours in the future. They are specific about the times that different weather conditions can be expected to change.

Forecasts are also prepared in a graphical form for the area in which aircraft are flying. These maps depict inflight weather conditions across the area and are used by aircraft pilots to infer weather conditions at airports where no TAF is issued. These Graphical Forecast Area charts are issued every six hours for six-hour intervals, out to 18 hours. The maps include the pressure pattern and frontal position forecasts, and they depict areas of cloud, including the expected height of cloud bases and tops, and areas where cloud bases will be near ground level. Areas of precipitation and regions where visibility near the ground might be obscured are also shown. Identifying regions in which aircraft icing conditions might be expected is important for safe aircraft operations, so these areas, including anticipated icing rates, are depicted. Atmospheric turbulence is a concern for passenger comfort and aircraft safety, so areas of turbulence including the degree of severity and levels in the atmosphere where turbulence can be expected are shown. The wind pattern at flight level is useful for aircraft navigation and planning. Maps of winds at several levels in the troposphere are generated from the numerical weather prediction models that provide guidance to the weather forecaster. Again, these products are issued for six-hour intervals, out to 24 hours.

The Graphical Forecast Area chart

Summer Severe Weather

The summer severe weather program usually focuses on convective weather and severe thunderstorms. The period of concern extends from late spring until the end of summer. When considering a convective weather forecast, the meteorologist is most concerned about the vertical thermal stability of the atmosphere. The development and movement of convective clouds is the forecasting issue. Where will they form, and will they grow explosively to produce heavy rain, hail, wind storms or even tornadoes? The severe weather forecaster needs to have a good three-dimensional picture of the atmosphere. This picture is built up using radiosonde observations, as well as analyses and predictions of temperature, moisture and wind patterns from the ground up to the tropopause. Images from weather satellites and weather radar provide valuable detail about the structure and motion of the weather systems and existing convective storms. The forecaster must be attuned to the influence of low level moisture from sources such as transpiring forests or croplands. Thunderstorms can develop and grow to severe intensity over a period of an hour or so, and they may move rapidly. Watches and warnings for specific sites need to be issued promptly for areas where severe thunderstorms are anticipated so that the public can be warned.

Particularly challenging are forecasts of tornadoes, which typically have a lifespan of a few minutes to one hour. Every severe thunderstorm is not a tornado producer. Even sophisticated observational

179

tools, such as the weather radar, do not have high enough resolution to see the small tornado funnel. However, radar can measure wind speed over larger scales using what is known as the Doppler shift technique. Tornadic thunderstorm cells are usually rotating, and the Doppler radar can frequently detect the large-scale rotation of a supercell thunderstorm. The regular observation network is too sparse to delineate small-scale features or observe small-scale events, so the severe weather forecaster depends on observations made by a network of voluntary severe weather watchers. Only when a tornado has been observed to touch the ground is a tornado warning issued. Time is of the essence in issuing the forecast and getting the warning to the public.

> A lot of people like snow. I find it to be an unnecessary freezing of water.
>
> –Carl Reiner

Winter Severe Weather

Winter severe weather includes heavy snowfall or rainfall, freezing precipitation, wind chill and strong winds. In British Columbia, these conditions occur from late autumn until spring. However, even in summer months there is a chance of winter-like conditions at higher elevation passes such as the Coquihalla Connector and the Salmo-Creston highways. Winter severe weather usually takes a longer period to develop than does summer severe weather. The large-scale prediction of weather patterns is heavily relied on to determine the location, intensity and motion of the winter storm event. The NWP products provide good guidance, but errors in the position of the features can make all the difference in forecasting precipitation rates or blizzard conditions, and careful attention is paid to pressure patterns and their evolution. Weather observations play a key role, and observations from volunteers

are relied on to fill in the details. The start of snowfall is monitored at observation stations, and the forecaster does subjective correlation between snowfall observations and the precipitation patterns displayed by weather radars.

Temperature and wind predictions are used to prepare wind chill forecasts. Areas of above-freezing temperature layers above the ground determine where freezing precipitation may occur. Freezing precipitation is usually difficult to forecast more than a few hours in advance—a problem that is similar to the summer severe weather forecast. Forecasters must anticipate the area where the event may happen and then use all observational data for clues to its occurrence and intensity.

Marine Forecasts

These forecasts are issued four times daily for the waterways along the BC coast. Wind directions and speeds are the main part of the forecast. When strong winds are expected, warnings are included for the various speed ranges, from strong through to hurricane force (see Figure 8-2). Precipitation is also mentioned, as is fog if visibility is expected to be less than 1 nautical mile.

The forecasts are general and may not always depict the wind conditions through the inlets and passageways at the coastal edge. Experienced mariners are aware of local wind and associated wave effects that happen with the various weather systems that affect the coast.

Add the tricky tidal effects, currents and marine hazards to the mix, and you can see why this section of the coastline is often called the "Graveyard of the Pacific." There have been so many shipwrecks over the years that it may be a low estimate to say that there is a wrecked ship for every nautical mile of coastline. Modern navigation systems and more accurate forecasts have helped to reduce the number of incidents, especially to larger craft. However, wrecks and drownings still occur every year, especially in the summer boating season when inexperienced boaters are caught by winds and waves beyond their capabilities.

The Role of the Forecaster

Meteorologists in the weather centre in Vancouver prepare all public and marine forecasts and warnings for British Columbia and the Yukon. Five or six forecasters are on duty every shift, and the team works together to make consistent decisions on the evolution of weather systems across the forecast region. The forecasters in a weather centre talk to their counterparts in the offices in adjacent regions to ensure that the weather systems are handled consistently as they move across the country.

The weather forecast program operates 24 hours per day every day of the year, so shift work is the reality for forecasting staff. The meteorologist plays a vital role in the weather forecasting process. Although observational technology and computerized forecasting tools are sophisticated, the weather forecasting process requires a human touch. The quick synthesis of data and the ability to recognize patterns and make decisions in the face of conflicting or incomplete data are tasks best left to the human forecaster.

The first task for a meteorologist starting a shift is to get a briefing from the staff members who are finishing a shift. Which weather systems are active in which sections of the forecast region? What forecasts and warnings have been issued? What is the confidence level that conditions will evolve as predicted? Today's weather office is moving toward a paperless working environment, and the computer-supported graphical workstation has replaced the map displays that were once the predominant feature. The meteorologist reviews observed conditions over the region of forecast responsibility and often undertakes a hand analysis of the computer-plotted surface weather map. This hand analysis is one of the few paper products still used because the process helps focus the meteorologist's attention on the details and occasional data anomalies and inconsistencies. Hourly weather data from automated and staffed observation sites across the forecast region is continuously streaming into the office through the communications system. Computers organize the data, and computer algorithms compare the observations with the valid forecast. If there is a significant discrepancy, the meteorologist is notified so that an amendment can be issued.

A large variety of other depictions of the atmosphere are generated. Weather patterns at various levels in the vertical, and graphs of the vertical temperature and moisture structure provide the meteorologist with knowledge of atmospheric conditions favourable for convection. Imagery from geostationary and polar-orbiting satellites is continuously displayed. Displays of weather radar images provide a time sequence of cloud patterns and precipitation rates across the province, and the lightning detection system indicates where thunderstorms are active. The meteorologist uses his or her knowledge of atmospheric processes and the influence of topographic and other controlling features to diagnose the current atmospheric condition and its short-term change.

> **Old weather saying:**
> *As high as the weeds grow,*
> *so will the bank of snow.*

The forecast products are all prepared at the computer workstation. The computer automatically generates a time sequence of all the weather elements, such as cloud, temperature, wind, precipitation and visibility, to be included in the forecast for all locations in the region. The meteorologist makes any necessary adjustments to the weather element sequence using an interactive knowledge-based system. When the forecaster is satisfied with the forecast, a command is sent to the computer system to generate the worded forecast, which is transmitted to media outlets. The product is also sent to a voice generation system that automatically prepares an audio clip for broadcast on the weather radio network across the province.

Weather forecasters hard at work (above and left)

For aviation, Terminal Aerodrome Forecasts (TAFs) are prepared for each of the larger airports across the province. Again, the computer guidance material is used, but the worded forecast is input directly into a bulletin by the meteorologist. Aviation forecasts include many weather parameters, and a continuous comparison is made between observations and forecasts. The comparison is largely performed by the computing system, which automatically sends out a warning to the forecaster if discrepancies between the forecast and observed values exceed predefined limits. The meteorologist must quickly evaluate the reasons for the change and issue an amendment to the TAF.

The forecaster also composes prognostic Graphical Forecast Area charts directly on an interactive aviation workstation. Again, the material generated by the central computer is used as a guide. The forecast surface pressure patterns are based on the NWP guidance. The forecaster adds frontal positions and outlines areas of cloud and precipitation, as well as regions where icing and turbulence are to be expected. All these fields are developed from the temperature, moisture and wind patterns in the NWP guidance.

Reading weather data

Chapter 11:
Fire Weather

For thousands of years, wildfires have raged each summer in British Columbia's forests. They probably date back to the end of the last ice age, about 10,000 years ago, after a lightning strike ignited the first growth of trees. Until recent years, wildfires were left to burn because First Nation peoples were unable to suppress them. However, with the sparse populations of the past, the chance then of a settlement being in the path of the spreading fire was a lot less likely than the threat to the dense network of cities and towns today.

Wildfires are considered a natural part of an area's ecosystem—part of the cycle that involves first the growth of forests, then a fire that develops and burns off some of the vegetation, and eventually regrowth. In a world with no people or structures to protect, letting each fire burn is the most ecologically sound procedure. The fire burns off the older, drier vegetation before it can accumulate to levels that would feed a forest inferno. In the natural fire cycle, some trees are destroyed, but the fire is mostly confined to the undergrowth. In the

years following a fire, fire-resistant seeds sprout and forest floor plants flourish, creating habitat for wildlife. Sometimes fires are deliberately set to mimic this cycle. In May 2008, the Parks and Protection Branch of the British Columbia Ministry of Forests conducted a prescribed burn in the Spatsizi Plateau Wilderness Park in northwestern BC. The purpose was to enhance Marion Zone Stone Sheep habitat by increasing grass production.

Fires that start near populated areas or in valuable timber are in a completely different category. As soon as these fires are spotted, fire crews mount an aggressive attack to extinguish them. Depending on the fires' behaviour, aircraft carrying retardant and heavy equipment such as bulldozers are used to control them. For close to 100 years, BC Forest Service fire fighters have been very successful in early detection and suppression of wildfires in the province. Soon after a fire is reported, three-person initial attack crews race to the location and put out many fires before they have a chance to spread. If the fire is beyond the crews' capabilities, helicopters and air tankers are immediately employed to drop water and retardant on the leading edge of the fire. In most cases, fires can be controlled in the first few hours after ignition. In August 2008, while I was working at the Castlegar Fire Centre, lightning strikes started over 100 new fires in one day. With the quick action of crews, only one fire spread to over 3 hectares!

Unfortunately, the incredible success rate of fire suppression has left a huge amount of forest floor litter that traditionally would have been burnt every decade or so. Every year the piles deepen, and when a fire breaks out in a very dry location, the litter acts as fuel and the flames soon leap up to tree level.

The fire can race almost instantly to the top of the tree in a process called candling. If it is windy, the fire spreads from tree top to tree top (called crowning), and soon there is very little that can be done to stop the progress of the fire. In the past 10 years, a number of these intense fires have destroyed houses and consumed thousands of hectares of trees.

Weather plays a huge role in the creation of a fire environment in forested areas. In fire sciences, three very important elements are associated with wildfires—weather, topography and fuel, with weather most influential. The weather patterns throughout the fire season determine the degree of dryness and volatility of fuels. In years with average summer temperatures and precipitation amounts,

fires are generally smaller and easier to contain. In years with extended dry spells and hot temperatures, such as 1998 and 2003, the fire danger rises to extreme, and fires can quickly turn into infernos.

There has been a significant rise in the number of bad fire seasons in British Columbia. While additional fuel is responsible for some of this increase, climate change has been targeted as another factor. Many summers in the past decade have been warmer and drier than years before, especially in southern BC, and almost every year there have been some large and damaging fires. Although 1998 and 2003 were extreme, other years have been notable for localized fire damage. Very little rain fell in southeastern BC in July and August in

2007, and temperatures soared to the low to mid-30s most days. Large fires burned throughout the summer, culminating with the Pend d'Oreille conflagration near the U.S. border in mid-August. However, fires aren't confined to the southern parts of the province. An evacuation order cleared the 3500 residents of Tumbler Ridge in the Rocky Mountain foothills in early July 2006 when a huge fire was less than 5 kilometres from the town.

Climate change has also been responsible for a large increase in the number of pests in the forests. With less extreme cold winter events, the number of bugs that survive winter has been rising. A period of at least five days of -20° C in autumn or -40° C in late winter is needed to kill the bugs, and such a cold spell has not occurred for many years. The mountain pine beetle infestation has reached epidemic proportions. Nearly half of the merchantable pine trees in the Southern and Central Interior of the province have already been killed. By 2013, that figure is expected to rise to 80 percent. These "red and dead" trees—those that show needle damage and eventually lose their needles altogether—add to the risk of fire starts, and fire behaviour with these trees is much worse than with live trees. With a projection of continued warming over the next 50 years, tree losses from insects can be expected in other types of conifers as well. Also, the number and severity of forest fires will almost certainly be on the rise.

Canadian Forest Fire Danger Rating System (CFFDRS)

The CFFDRS is a national system that rates the ease of ignition and the difficulty of controlling forest fires in Canada. Risk factors (lightning and human-caused fire starts), fuels, topography and weather provide the necessary information to predict fire weather, fire occurrence and fire behaviour. To produce a Fire Weather Index (FWI), various weather elements are loaded into the system. The FWI is a general index used to rate the fire danger in forested areas of Canada. As you drive around the province, you will see BC Ministry of Forests signs displayed at the side of the road with categories ranging from "very low" to "extreme." Figure 11-1 shows some other indices that use weather data to predict the dryness of the forest floor litter, initial spread potential after ignition and the total amount of fuel available for combustion.

The Protection Branch of BC Forest Services has been responding to wildfires in the province since its inception in 1912. Highly trained crews, which are internationally recognized as leaders in wildfire management, respond to an average of 2000 wildfires each year. Approximately 92 percent of all wildfires are contained within 24 hours of discovery. Interface fires where communities are threatened have the highest priority. In the years to come, as cities continue to spread into wildland areas, these fires will be the biggest challenge for the Protection Branch.

Fire Rating Indices

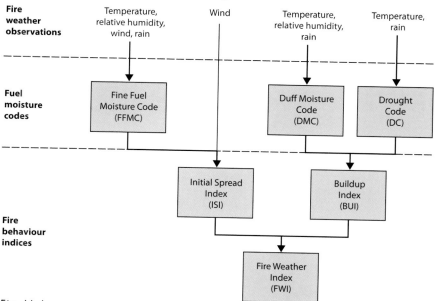

Fig. 11-1

189

Fire Weather System

BC Forest Services' Protection Branch maintains a network of 215 automated weather stations across the province. Weather networks from other agencies such as Environment Canada are used for additional information. Every hour during the fire season, temperature, relative humidity, wind velocity and precipitation data is transmitted to the Protection Branch headquarters in Victoria. The readings taken at noon Pacific Standard Time are used to update fire weather indices, which give an indication of the trend in forest dryness and the volatility of fuels.

The daily figures measured by the weather stations are not enough to provide a complete picture of fire risk. Fire weather meteorologists prepare forecasts for every fire centre in the province. Weather data is predicted up to one week ahead using selected stations within the weather network. The forecasters also brief forestry personnel

each morning using weather maps and computer projections. All weather elements are covered, but special attention is given to the chance of lightning when the risk of fire in the province is increasing. When fires are already burning, fire behaviour is very important for the safety of the public and fire crews. Any suspected hot temperatures, low humidities and especially strong winds are emphasized to keep people out of harm's way and to protect structures and valuable assets such as timber and grasslands.

Based on actual and forecasted numbers, decisions are made regarding the allocation of resources, both human and equipment. Crews and aircraft are stationed around the province in areas where fires are burning with the most intensity, where there is a risk of severe fire behaviour or where there is the highest risk of fire starts. Firefighters are often deployed to other provinces and countries when the fire danger is low in BC. In winter, management teams have even been sent to California and Australia to help with their fire seasons.

The 2003 Firestorm

The 2003 fire season smashed all the previous records for the total area of forest burned (260,000 hectares), structures destroyed (334 homes and many businesses) and firefighting costs and property losses ($700 million). More than 45,000 people were evacuated. The greatest cost, however, was the loss of lives—three pilots died in the line of duty.

For the previous three years, the weather had played a huge part in creating the perfect environment for wildfires. Ten out of the 12 previous seasons had been drier than normal in the southern third of BC, and most months were warmer than average. The forest floor litter all the way down to the deepest layers, called duff, had much less moisture content than usual, coming out of a winter with below average snowpack. However, there was no cause for alarm in late spring because the rain amounts through May and until the third week of June were only slightly below average. Then the drought set in. From June 23 until September 10, a span of 80 days, less than 5 millimetres of rain fell at both Kamloops and Kelowna. No measurable rain fell on Kelowna for a record 44 days.

Temperatures were also well above normal, especially in July, with maximum temperatures from the Chilcotin to the U.S. border averaging over 30° C, up to 4° above normal. Kamloops and Castlegar soared to 40° C on July 31.

The first major fires developed in the southern Chilcotin during the third week of July, near Tatla and Chilco Lakes.

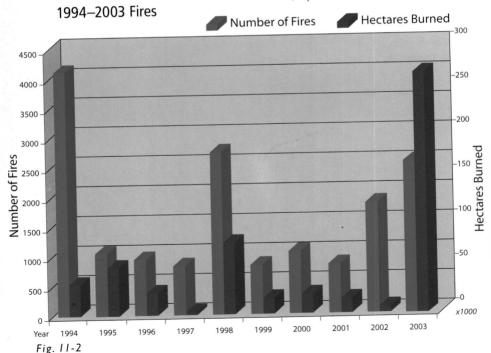

1994–2003 Fires

Fig. 11-2

The fires showed little activity until a dry cold front with gusty winds passed through, and the fire behaviour immediately rose to extreme. The Chilco fire expanded to over 29,000 hectares by early August. Fortunately there were few settlements in the area, but many animals were threatened, including a herd of wild horses near the lake at the Brittany Triangle in the Nemaiah Valley.

The number of interface fires (those affecting urban areas) accelerated from the end of July through to the third week of August. The list of fire locations is long—McClure, Strawberry Hill, Cedar Hill, Mcgillvray, Lamb Creek, Okanagan Mountain Park and Vaseux Lake, to name a few. Figure 11-3 shows a list of notable fires and the communities they affected. The McClure and Okanagan Mountain Park fires were the most devastating.

Major Interface Fires, Summer 2003

Start Date	Fire Name	Location	Size (in ha)
July 22	Chilko Fire	Chilko Lake, Alexis Creek	29,202
July 31	McLure Fire	McLure, Barriere	26,420
August 16	Okanagan Mountain Park Fire	Kelowna	25,600
August 1	McGillivray Fire	Chase	11,400
August 16	Lamb Creek Fire	Cranbrook	10,979
August 16	Venables Fire	Chase	7636
August 17	Ingersol Fire	SW of Nakusp	6700
August 1	Strawberry Hills Fire	Kamloops IR	5731
August 20	Kuskanook Fire	N of Creston	4832
August 22	Vaseaux Fire	Okanagan Falls	3300
August 14	Plumbob Fire	Cranbrook	2870
August 2	Cedar Hills Fire	Falkland	1620
August 6	Bonaparte Lake Fire	Bonaparte Lake	1500
August 17	Anarchist Mountain Fire	Osoyoos	1230
August 20	Harrogate Fire	Radium	1018

Fig. 11-3

The McClure fire was started by human carelessness, and it damaged or destroyed 72 homes and nine businesses including a large local employer, Louis Creek sawmill. At its largest, the fire covered more than 26,000 hectares; 3800 people had to be evacuated.

The Okanagan Mountain Park was the most significant interface wildfire in BC history. Started by lightning on August 16 near Rattlesnake Island, the blaze destroyed or damaged 239 homes, forced the evacuation of 33,000 people and claimed 12 wooden train trestles in the historic Myra Canyon. Sixty fire departments, over 1000 forestry firefighters, contractors and loggers, and 1400 members of the Canadian Forces fought the fire from mid-August until the end of September. On August 22, the fire reached its peak of destruction. Most of the homes were lost that day, and an evacuation alert was posted for a huge area because there was a strong possibility that most of Kelowna and areas all the way north to Winfield might be affected. Miraculously, a short but intense evening thundershower hit the head of the fire and stopped the fire's progress in its tracks; otherwise, many more homes would have been lost.

As the spread of houses continues into forested areas and the number of insect-damaged trees increases, the danger of fires each summer, especially those affecting communities, increases as well. Within the next few years, BC will most likely face another destructive season like that of 2003. The trigger for starting the blazes will always be the weather. One area of the province that has seen fewer fires in the last decade is the northern half. The weather in the north is generally wetter, but the region has had dry seasons in the past and is likely to experience another one eventually. With climate change studies predicting a general rise in temperatures in the future, more numerous and intense fires are possible throughout the province.

Chapter 12: Air Quality

British Columbia has a diversified economy and a growing population base. Although population growth in BC's cities is more rapid than in the rural areas, air pollution sources span the province.

Emissions to the Atmosphere

Activity associated with oil and gas exploration, production and processing is the largest source of atmospheric emissions. Agriculture, home heating, transportation and electric power generation are also major emission sources. Pollutants are classed as either primary or secondary. Primary pollutants have well-defined sources such as smokestacks and vehicle tailpipes. Secondary pollutants are those that are the product of chemical reactions in the atmosphere. These reactions occur among the primary pollutants and with various constituent gases in the clean atmosphere.

The high temperature combustion of fuels produces most of the emissions of concern. Nitrogen oxides are produced when any combustion occurs in the oxygen and nitrogen gas mix of the atmosphere. They may also be produced by the oxidation of nitrogenous compounds that are in the fuel. Nitric oxide (NO) and nitrogen dioxide (NO_2)—collectively known as NO_X—and nitrous oxide (N_2O) are the most important nitrogen pollutants. Nitrous oxide is also significant because of its radiative properties. It strongly absorbs infrared radiation and is one of the main greenhouse gases. Stationary sources (power plants and residential and industrial heating) and mobile sources (vehicles and aircraft) contribute, and BC accounts for a significant amount of Canada's NO_X emissions. Carbon monoxide (CO) and carbon dioxide (CO_2) are also products of combustion and are formed when the carbon in fossil fuels is oxidized by atmospheric oxygen. Many fossil fuels contain trace amounts of sulphur and, though much of the sulphur is extracted from fuels, residual amounts are converted to sulphur dioxide (SO_2) during combustion. Hydrogen sulphide (H_2S) is naturally present in natural gas and oil reservoirs. It is a toxic gas with a characteristic rotten egg smell. In the production of fossil fuels, H_2S is extracted and converted to elemental sulphur, but small amounts of residual H_2S are released or transformed into SO_2 through combustion.

A wide variety of hydrocarbons find their way into the atmosphere. These volatile organic compounds (VOCs) originate from the evaporation and combustion of liquid fossil fuels and from chemical manufacturing, petroleum production and solvent use in chemical processes and domestic applications.

Ammonia (NH_3) plays a role in the production of particulate matter (PM) in the atmosphere. Animal husbandry, fertilizer application and some industrial processes are sources of ammonia in BC.

Atmospheric particulates are contaminants that are attracting increasing concern. Some particulates are classed as primary pollutants, but many particulates are secondary pollutants that form in the atmosphere. Particulate matter impairs the clarity of the atmosphere by degrading its light transmission properties, resulting in reduced visibility of distant objects. It can also have negative health impacts. Particulates are generally categorized according to their aerodynamic size based on their ability to be deposited in lungs. Larger particles usu-ally have a natural origin, such as wind-blown dust from dry soil or re-suspended dust from road surfaces, but they can also originate from combustion as smoke, soot and ash. These particles are generally larger than 10 microns in diameter. It is the smaller particles—generally with a size less than about 2.5 microns—that find their way into the lungs and cause the greatest health risk. These particles may be either solid or liquid droplets and are often formed from the gases that are a product of combustion.

Other compounds that are in the categories of toxics, heavy metals and persistent organic pollutants are either minor by-products of industrial processes or are heavily regulated, and few reach the atmosphere in BC.

Emissions of pollutants vary greatly across the province. With their high population concentrations and intense levels of industrial activity, urban centres and their surrounding areas emit the majority of pollutants. Agricultural emissions are widespread in the cultivated lands in the Fraser Valley and BC Peace, with lesser amounts in other parts of the province. Processing facilities for oil and gas extraction release pollutants in the northeastern portion of the province. Industries report their annual emissions to the government of Canada for archiving in the National Pollution Release Inventory.

There is an atmospheric cycle through which pollutants are mixed, moved around, sometimes transformed and finally removed from the atmosphere by several natural processes. Polluting gases

Slash burn

are diluted downwind of the emission sources. Although the saying "the solution to pollution is dilution" was often heard, dilution is taxing on the clean air resource. Consider vehicle emissions of carbon monoxide. In 1965, the average emission rate was 30 grams of CO per kilometre of travel. About 2 million litres of clean air—the amount breathed daily by 200 people—are required to dilute that much CO to levels that are safe for humans to breathe. So, it takes large volumes of air to dilute all the pollutants to levels that are considered safe. In many regions of the world, there is not enough clean air, and an emphasis was placed on improving vehicle efficiency along with reducing pollutant emission rates. By 1997, the emission standard for CO was reduced to 2.1 grams per kilometre. On a per vehicle basis, we are now polluting significantly less air.

Cloud condensation nuclei are in part composed of the small particles originating from pollutants. Studies of rainfall patterns have revealed how pollution in the atmosphere influences precipitation. For example, enhanced rainfall within Chicago relative to its less polluted suburbs has been attributed to the vast number of condensation nuclei injected into the atmosphere by industrialized regions in that city. Farther downwind, enhanced rainfall in the city of LaPorte, Indiana, has been attributed to the transport of nuclei from the same upwind Chicago pollution sources. The rainfall process is also a cleansing mechanism for polluted atmospheres—it removes and washes away some nuclei. In anticipation of the 2008 Olympics, China artificially enhanced precipitation in Beijing to rain out the heavy pollution concentrations and improve air quality.

Pollution Movement in the Planetary Boundary Layer

The lowest layer of the atmosphere is known as the planetary boundary layer (PBL). Within the PBL, most life forms are subjected to air pollution. Air quality and the dispersion of pollutants are influenced by the turbulent motion and temperature structure within the PBL. Three key factors determine the characteristics of the PBL:

1. The transport of heat from the earth's surface. Heating caused by the solar radiation cycle, heating from phase changes of water (latent heat) and the spatial variation in heating (near lakes and over different land surfaces) are all examples of the complex heating process.

Heat is an important energy source that leads to mixing of the PBL—just like the heating element on a stove causes water in a pot to turn over and mix.

2. The influence of roughness at the earth's surface. Airflow is disturbed by the presence of objects, from blades of grass or trees to buildings, hills and mountains. These objects result in a surface roughness that slows the airflow in the PBL by imposing what is known as frictional drag. Large objects can also deflect and block airflow. Roughness features induce mechanical turbulence, much like water flowing over a rough riverbed is turbulent. The larger the roughness features and the stronger the speed of the flow, the greater the degree of turbulence.

3. Airflow and thermal structure of the atmosphere at the top of the PBL. This level is the bottom of what is known as the free atmosphere and is characterized by the atmospheric flow associated with large-scale pressure patterns. At this level, the geostrophic wind law describes the horizontal flow. Upward and downward motions associated with the large-scale flow are also evident. The top of the PBL can be a sharp transition zone into the free atmosphere. If the temperature in the PBL is lower than in the immediately adjacent free atmosphere, the top of the PBL effectively acts as a lid that can prevent the upward motion of pollutants out of the PBL.

Turbulence

Atmospheric motion is anything but uniform—it is turbulent. Whether winds are light or strong, we sense the turbulence as gustiness. When wind measurements are recorded, the time series shows frequent changes in wind speed and direction. These fluctuations are characteristic of this turbulent flow. Air quality scientists think of wind as having two components. The first component we can call the wind flow. It is considered to be steady for some interval of time. The second component is the perturbation, which is the slight departure from the mean wind. The motion of air pollutants is governed by the wind flow itself and by perturbations in that flow. Wind flow is a vector quantity—it has speed and direction, and it is governed by the mechanical equations of motion of the atmospheric fluid. The mathematics dealing with perturbations in the wind, known as the turbulent component of the airflow, are complex.

An old saying that demonstrates the difficulty of understanding turbulence goes, "when you get to heaven, God will offer two explanations—one for quantum physics, and the second for turbulence theory." Air quality science focuses on the effects of such turbulent motion.

The rates at which pollutants disperse and how pollution concentrations change with time are described by the mathematics of turbulence theory. This theory can be put in the form of mathematical equations, which are used to calculate the rate of spread of pollutants and the changes in pollutant concentration downwind of an emission source.

When pollutants are emitted into the atmosphere from a point source, such as a smokestack, they have two key characteristics. If they are warmer than the surrounding air, their buoyancy causes the pollutants to rise. They also immediately start mixing because of turbulence and spread out laterally from the source. If a wind is associated with the fluid, the pollutant cloud takes the form of a plume. The plume has a characteristic shape, as shown in Figure 12-1. It initially moves vertically then is bent over by the wind. With time, the polluting gas spreads away from the centre line, taking on a conical form. The shape of the cone, and the pollutant concentration within the cone, is described by the statistical properties of the turbulent atmosphere.

Temperature Structure

The temperature structure of the planetary boundary layer has an important influence on air quality because the vertical temperature structure controls the vertical mixing. As mentioned earlier, the rate of change of temperature with height is known as the lapse rate. The lapse rate is usually negative—that is, temperatures decrease with increasing height. The greater the rate of temperature decrease with height, the greater the ability of the atmosphere to disperse pollutants. Most pollutants are emitted near the ground, so the lapse rate in the PBL is important in dispersing pollutants. However, there are many situations in which temperatures increase with

Depiction of a Plume Rising in a Stable Atmosphere

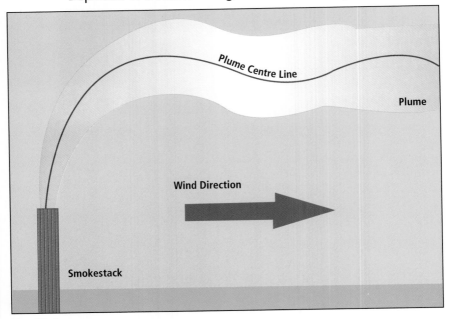

Fig. 12-1

height, creating an inversion. An inversion condition in the atmosphere slows the dispersion process, and pollutant concentrations can continue to increase near a source that is continuously emitting.

The lapse rate varies widely during a 24-hour period. When the sun sets, the ground surface cools, as does the air in contact with the ground. Air farther aloft does not cool as quickly, and a nocturnal inversion develops. Pollutants emitted into this inversion layer may move laterally with the wind, but they have little vertical movement. The nocturnal inversion often traps pollutants near the ground.

Other types of inversions are also important in BC. The interface between two different air masses is known as a front. Frontal inversions can play a role in trapping pollutants near the ground. The frontal surface can have a shallow slope, and a large area of the province can be subjected to frontal inversion effects, especially in the cold seasons with slowly moving warm fronts. Cold fronts, on the other hand, often move rapidly, and they clean out stagnant polluted air masses.

Terrain can also play a major role. Cooler air pools in river valleys and may be overrun by warmer air. The resulting inversion can trap pollutants within a valley for days at a time.

In December 1952, more than 4000 people in London, England, died as a result of a severe pollution event. A strong thermal inversion acted as a lid and trapped moisture and pollutants in the PBL above the city. Incident sunlight was reflected by higher level cloud decks and by the top of the fog layer itself. Sulphur-laden smoke from coal-fired furnaces and stoves accumulated over a period of five days as the inversion persisted. Most people died because of respiratory tract infections, such as bronchitis and bronchopneumonia. Largely because of this event, the term smog—meaning a mixture of smoke and fog—was coined.

British Columbia never experiences such dramatic events. However, a combination of pollutants, especially ground-level ozone, is a significant problem in summer months in the Lower Fraser Valley. Typically the afternoon westerlies help to clear the smog around the city of Vancouver and send it eastward toward the Coast Mountains. A "damming" effect occurs as the air reaches the higher ground, especially with stable conditions under an upper ridge of high pressure. Hope and Chilliwack have their worst air quality readings under these conditions.

London, England, used to be known for its so-called pea-soupers: heavy blankets of smog, made up of a combination of fog and smoke from factory smokestacks and the burning of coal. In December 1952, 4000 people died of chest disease as a result of pea-soupers.

Formation of Photochemical Smog

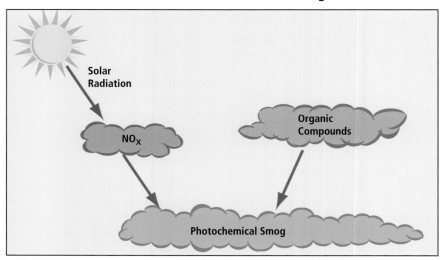

Fig. 12-2

Chemical Transformation in the Atmosphere

Photochemical smog is common in all large cities where vehicles, industries and other pollution sources are concentrated. Most prevalent and well known were the smog events that frequented greater Los Angeles. Through the 1960s and '70s, the causative chemicals and chemical reactions responsible for the smog were determined, and a solution to the issue has progressed with California's imposition of stringent vehicle emission controls. Photochemical smog is now a much less frequent occurrence, and events are less severe than during its peak in the 1970s.

Photochemical smog consists of a mixture of the primary emissions of NO_X and VOC gases, and the secondary pollutants ozone (O_3) and particulates.

These two secondary pollutants are produced when intense sunlight acts on the mixture of primary emission gases. Reactions in the polluted mixture proceed depending on the concentrations of the precursor gases, the temperature and the amount of mixing that is present in the planetary boundary layer. Some of the reactions are rapid, some are slow and some depend on the intensity of sunlight. Generally, the reactions in the polluted soup proceed to what is known as an equilibrium state. In the worst-case scenario, this equilibrium state has high concentrations of ozone and particulates.

British Columbia's major cities have significant emissions of the pollutants that can lead to the production of photochemical smog. The presence of NO_2 is evident because this gas preferentially absorbs light in the short wavelengths (blue to green). When light passes through the gas, these short wavelengths are absorbed, allowing the longer wavelengths—the red to orange colours—to pass through the gas. As a result, the gas appears to have a brownish colour. The brownish tinge is usually evident downwind of Vancouver during light wind conditions in summer and winter. Meteorological conditions in summer (calm winds and intense sunlight) favour the production of photochemical smog. Because the reactions can take a few hours, peaks in concentrations of ozone and small particulates vary in space and with time. Air quality monitors near downtown Vancouver usually show elevated NO_x concentrations, especially during the morning rush hour when there is more NO_x available than is necessary for ozone production. Ozone reaches its highest level late in the afternoon when solar intensity reaches a peak downwind of the city centre over the middle and eastern Lower Fraser Valley. Overnight, concentration levels of most of the primary and secondary gases drop to a minimum.

Smog situations can also occur in winter. They characteristically have high concentrations of PM and NO_x, but ozone levels are low because of low temperatures and weakened solar intensity. Winter smog usually forms because of the presence of strong thermal inversions and low wind speeds that trap pollutants near the ground. Chemical reactions proceed slowly at low temperatures, though low temperature conditions are favourable for the formation of some types of particulates. Interior valleys often have the worst concentrations of chemicals, especially particulates from sawmills and pulp mills. Stagnant pollution trapping events are more common in late autumn in Vancouver.

Air quality monitoring station

forms of the acids can be deposited in dry form, while the acids in aqueous form are deposited in precipitation known as acid rain.

Negative impacts of acid deposition usually depend on the nature of the receptor. Acids can be neutralized or rendered less harmful by some chemicals that occur naturally in many receptors. However, the groundcover and lakes in areas of Precambrian rock (the shield areas of northeastern BC) have low buffering capacity and can be susceptible to acidification. As well, because the acids and the chemicals that are acid precursors can be transported long distances by the winds, acid deposition can occur far from the pollution source. This transport is the reason for much of the acidification of sensitive lakes in eastern Canada and the United States. Many of the polluting sources are well upwind.

Forest fires can also have a major influence on air quality in BC. The smoke, of course, has a high particulate loading and itself can be responsible for health distress. Reports of associated illness and increased admissions to medical centres occur at locations well removed from the fires. Chemical reactions also occur in the smoke plume. Many of the reactions and chemicals are the same ones involved in the photochemical smog problem. Smoke plumes from forest fires contain elevated levels of NO_2 and VOCs produced from the burning of massive quantities of biomass. Some of the highest levels of ozone and particulate matter have been measured at sites beneath smoke from forest fires when ozone and small particulates are mixed down to ground level.

The formation of acids in the atmosphere and the deposition of these acids on lakes and forests is an issue in much of eastern North America and Europe. Sulphur dioxide and NO_x both play a role, and the high rates of emissions of these gases in BC have been the focus for a considerable research effort. Sulphur dioxide has been the primary culprit in the acid rain issue. It transforms into sulphuric acid in the presence of water vapour and other atmospheric chemicals known as free radicals. Nitric acid (HNO_3) forms in the atmosphere as a by-product of the same chemical reactions involving NO and NO_2 that are responsible for ozone production. The gaseous

Air Quality as a Societal Issue in British Columbia

Where maintaining a healthy environment is concerned, governments usually assume the lead in establishing criteria to protect air quality and human health. British Columbia has established air quality objectives for many air contaminants, and these objectives are meshed with national goals. They are expressed in pollutant concentration levels in the atmosphere. Averaging times for pollutant measurements are specified, usually as hourly averages, though daily and annual averages are also used. These averaging times are closely linked to impacts—for example, how much time a plant, animal or human will be exposed to the pollutant.

In addition to the ambient air quality objectives, governments also establish regulations on emissions from various sources of pollutants. For example, the federal government has enacted regulations on emissions from cars and trucks, aimed at reducing the levels of NO_x in cities. At the same time, the provincial government has enacted regulations to reduce emissions from sources such as coal-fired power plants and from flares operated by the oil and gas industry.

In British Columbia, the government, industry and the public all have a role to play in dealing with the air quality issue. Several processes are in place to ensure that the three groups of stakeholders undertake joint discussions on many aspects of air quality. Areas for consideration include the setting of air quality objectives, the management of local air quality concerns, and approval of applications for new facilities that may have sources of air pollutants. The groups also evaluate the state of scientific knowledge on many aspects of provincial air quality, and if it is deemed insufficient, they may propose that detailed investigations and research be undertaken.

Chapter 13:
The Changing Atmosphere

The atmosphere includes a number of gases that have concentrations at trace levels. Some of the gases have a short residence time in the atmosphere, but some are long lived. Many of these trace gases play an important role because of their physical characteristics or because of their ability to participate in chemical reactions. These gases are important to the earth's environment and, indeed, to the health of the earth's ecosystem. One of the more significant groups of these gases is known as the radiatively active gases, so called because of the role they play in interacting with electromagnetic radiation—either the radiation emitted by the sun, or by the earth itself. Some of the gases have a natural origin, and some are only present because people release them into the atmosphere. Ozone and the greenhouse gases play a particularly important role for our planet.

Stratospheric Ozone

Ozone is a form of oxygen gas that has a dual role in the atmosphere—it is in some ways beneficial and in some ways detrimental. Ozone gas contains three oxygen atoms and is present at low concentrations throughout the atmospheric vertical column from the surface to the

Formation of Stratospheric Ozone

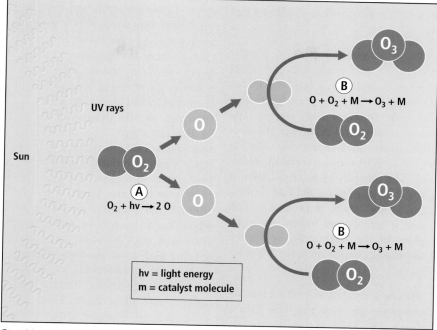

Fig. 13-1
A - ultraviolet light strikes oxygen, forming atomic oxygen
B - ozone molecules form in the presence of a catalyst molecule

stratosphere. Ozone in the stratosphere is produced through natural processes. The ozone molecules absorb much of the short-wave ultraviolet radiation emitted by the sun, protecting life forms from the damaging effects of ultraviolet solar radiation. This property of ozone has been known since the 19th century. Some ozone in the lower atmosphere is produced through the chemical reactions between gases that have natural and anthropogenic origins. This low-level ozone is detrimental to human health and the ecosystem, essentially because ozone is reactive with other substances. Ozone also plays a small role as a greenhouse gas. In the lower atmosphere, ozone contributes to warming of the atmosphere, while stratospheric ozone has a cooling effect. The phrase "good up high, bad nearby" describes the dual role of ozone.

Formation of Stratospheric Ozone

In the upper atmosphere, ultraviolet solar radiation at short wavelengths causes the dissociation of some oxygen molecules into atomic oxygen. The atomic oxygen reacts quickly with oxygen molecules to form the ozone molecule. This reaction also involves a third gas, usually

The Changing Atmosphere

the ever-abundant nitrogen. Nitrogen itself is unaffected, except that it absorbs some of the energy released in the formation of ozone.

The ozone molecule, in turn, absorbs ultraviolet solar radiation and splits apart to form diatomic (O_2) and atomic (O) oxygen. This cycle, in which ultraviolet light interacts with oxygen and ozone, is in natural balance in the mid-stratosphere and effectively absorbs harmful short-wave ultraviolet radiation. The intensity of solar radiation is greatest at the equator, and that is where most of the ozone is produced. This ozone is transported toward the polar regions by the large scale, or Hadley, atmospheric circulation pattern (see pp. 16–17).

However, substances emitted into the atmosphere are upsetting the natural balance. The gases of concern are the nitrogen oxides, formed in the combustion process, and other gases including chlorofluorocarbons (CFCs), which contain the chlorine atom, and chemicals containing the bromine atom. Nitrogen oxides (NO_x) emitted by combustion in the lower atmosphere normally undergo chemical reactions fairly quickly and do not find their way into the stratosphere, so the concern is with NO_x emitted at high altitudes. Although there were concerns over NO_x emissions from the fleet of supersonic aircraft proposed in the 1970s, there are relatively few aircraft flying at these critical heights, and the resultant ozone depletion from that source has been relatively small to date.

Of much greater concern is the role of CFCs and bromine compounds. When

CFCs were first discovered in the early 20th century, they were much valued because they are stable and don't react with other substances. CFCs were used as propellants in aerosol spray cans, and they replaced the dangerous ammonia gas in most home refrigeration systems. Chemicals containing bromine include methyl bromide, used as an agricultural fumigant, and the halons that are used in fire retardants. The problem with these gases is that they have a long lifespan. When introduced into the lower troposphere, these gases eventually work their way into the upper atmosphere through the action of thunderstorms, which are frequently so vigorous that their tops push into the lower stratosphere. The most energetic thunderstorms are at equatorial latitudes, so most of the penetration into the stratosphere takes place there. These long-lived gases are transported farther into the mid-stratosphere and eventually toward the polar regions by the same Hadley circulation that moves ozone toward the poles.

High energy ultraviolet solar radiation that enters the stratosphere breaks the chemical bonds in the stable compounds, releasing chlorine and bromine atoms into the stratospheric ozone layer. These atoms undergo reactions, forming what are known as catalyst molecules. The catalyst molecules react with the ozone molecules, breaking the ozone apart while the catalyst remains intact. The catalyst goes on to break apart another ozone molecule, and so on. Each chlorine or bromine catalyst molecule can effectively destroy thousands of ozone molecules.

The catalyst reactions that dissociate ozone do not proceed at a uniform rate or equally across the stratosphere. The most pronounced effects are in the polar regions where, during the cold season, stratospheric temperatures can drop to below -80° C. At these temperatures, polar stratospheric clouds containing a mixture of ice, nitric acid and sulphuric acid together with the catalyst chemicals are formed. A reservoir of catalyst chemicals is established in the polar stratospheric cloud layer. The effect is pronounced over the South Pole, where the circulation is effectively concentric around the pole and temperatures can reach the -80° C values. As the earth proceeds in its orbit and polar regions are turned toward the sun, the reservoir warms and the depleting chemicals are released. Destruction of ozone proceeds rapidly within and adjacent to the reservoir region. The circulation pattern also weakens around the reservoir, allowing ozone-depleted air to mix with ozone-enriched air that is moving poleward from the equator. In the northern hemisphere, with its greater land mass, the circulation pattern has a much greater degree of irregularity than in the southern hemisphere, and temperatures are not as extreme as those over the South Pole. For this reason, there is less ozone destruction in the north polar region than in the south.

211

Increased levels of UV radiation can expose us to a greater chance of skin damage from sunburn.

The rapid rate of thinning of the ozone layer was first noticed in the south polar region in the mid-1970s. The reduced ozone concentration has become known as the ozone hole, though it is not actually a hole; it is a zone where ozone concentrations are up to 40 percent lower than normal values. These lower concentrations allow much more of the damaging ultraviolet radiation to reach the earth's surface. Because of the increased health risks of solar ultraviolet radiation, governments agreed to take steps to deal with the ozone depletion issue. In 1979, many countries, including Canada, banned the use of CFCs as aerosol propellants. In 1987, an agreement known as the "Montreal Protocol on Substances that Deplete the Ozone Layer" was signed. The agreement required participating countries to reduce their produc-tion of CFCs. The agreement was later strengthened to phase out all CFC production.

Actions taken under the Montreal protocol are having a positive effect. Measurements in the high atmosphere now show that the concentrations of some of the ozone-depleting substances are slowly starting to decrease. There is, however, a long way to go. It will take until the middle of the 21st century for chlorine levels to return to the values measured in the 1970s when the south polar ozone hole first started appearing. Current projections by Canadian scientists indicate that global total ozone levels will return to 1960s averages by about 2060. The recovery assumes that countries will continue to adhere to the terms of the Montreal protocol and that no other long-lived substances will be found to cause ozone depletion. The possibility of increased NO_x emissions is still a concern because consideration is being given to allowing more high-flying supersonic transport aircraft. Another concern is the potential effect that global warming may have on stratospheric ozone in the polar regions. This is an ongoing topic of Canadian research.

Stratospheric Ozone and British Columbia

In British Columbia, the thinning ozone layer will continue to be a concern. In late winter to early summer, when ozone thinning is most noticeable over the province, the public is vulnerable to increased levels of ultraviolet radiation. At that time of year, the need to take extra protective measures is especially important for people who enjoy spring skiing. The intensity of UV radiation at alpine levels is enhanced by the thinner atmosphere and by the additional solar reflection off snow surfaces.

Climate Change

People in BC are well aware of the changes in weather from day to day. We know what to expect from our climate from year to year, but we also know that there will be some degree of variability. Some winters have more snow, and some summers are hotter or have more rain than others. Some years seem to have more severe summer storms, and some years are quite placid. This wide range of variability seems to be the norm. When averages of temperature, rainfall or any of the other atmospheric parameters are calculated at an observing station or are averaged over a region, much of the variability disappears. However, it is apparent from these averages is that there is an upward trend in temperature with time.

Is the climatic average moving to some new state, and if so, why? What will this new state mean for our lifestyle, for the environment? Can we change the trend? What measures will society need to take to deal with the change? What will be the cost of these measures? These are the many questions that are raised by the climate change issue.

Temperature is the most useful parameter to use in studies of climate and its changes. Temperatures are reasonably consistent over an area, and the temperature record can usually be readily established. The instrumented record only goes back about 130 years in BC. It is somewhat longer at other locations across Canada (back to about 1840 in Toronto) and only marginally longer at some sites around the world. Prior to the instrumented record, what can we say about the temperature and climatic conditions? Scientists have developed a number of techniques to deduce past conditions. We have anecdotal information about storms, cropping practices and settlement patterns that allow us to infer what temperature conditions were like over the last few centuries. From the examination of tree ring widths, we have reasonably direct evidence of the length and productivity of growing seasons as far back as several hundred years. The analysis of lake and ocean bed deposits provides a wealth of information about plant forms that were present on adjacent

People are finding ways to protect themselves from harmful sun overexposure.

lands or of plants and animals that inhabited the water bodies. This evidence from around the world helps push the climatic record back several thousand years. Carbon dating is a sophisticated tool that is used to look back about 40,000 years, and it is accurate to within about 100 years. The characteristics of ice in glaciers can also be used to infer what temperatures were like at the time the snow fell on glacier surfaces. Ice cores extracted from Antarctica and Greenland glaciers have been used to determine what temperatures have been like over the past 600,000 years.

The climate has undergone many natural transitions in the past. Northeastern BC was once covered with lush tropical growth, which is why we now have such a wealth of gas and oil reserves. Ice sheets have spread across the North American continent with a fair degree of regularity, every 100,000 years or so. There are several theories that describe why past climatic changes occurred. The first theory relates to the all-important link between the earth and the sun. The earth travels around the sun in a slightly elliptical orbit, and the earth's axis of rotation is tilted at an angle of 23 degrees. The Serbian mathematician Milutin Milankovitch examined the regular changes in the degree of orbit ellipticity, rates of change in the axis tilt and changes in the time of year that the earth is closest to the sun. With this information, he developed a schedule that could explain the onset and termination of past ice ages.

Shorter-term fluctuations in solar energy output are well known. The sun goes through a cycle in which sunspots come and go every 11 years. When the number of sunspots is at a maximum, there is a slight elevation in the rate of solar energy output. The increase in radiation is only slightly above the normal levels, though, amounting to less than 0.2 percent over the 11-year sunspot cycle. The sun's energy output may have deviated for periods of a half-century or so when there was an almost total absence of sunspots. The associated solar output decrease has been linked with a cold period in Europe in the 16th and 17th centuries, known as the Little Ice Age.

Earth's Energy Balance and the Greenhouse Effect

The temperature of the atmosphere is linked to the balance between incoming solar radiation and the properties of the earth's surface and the surrounding atmospheric envelope. Structures, water and life forms on the earth's surface absorb

Ice Age Temperatures—Greenland and Antarctica

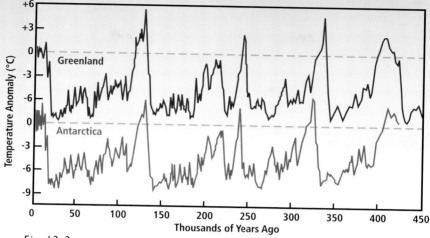

Fig. 13-2

a large fraction of the total incoming solar radiation. Objects that absorb radiation are warmed, and they in turn radiate energy in the infrared wavelength. Much of this infrared radiation is directly absorbed by radiatively active gases, which then warm and in turn radiate energy. All this energy transfer, absorption and retransmission are in a general state of balance over the earth. This balance is maintained by the meridional circulation. The global energy balance results in an annual average temperature of about 16° C over the entire planet. Of key importance is the role of the radiatively active gases. In 1824, famous French physicist and mathematician Joseph Fourier first proposed the concept that gases in the atmosphere trap heat from the sun. This process has been termed the greenhouse effect, though that is something of a misnomer—greenhouses prevent the mechanical movement of heated air, while

atmospheric gases reduce the radiative loss of heat. It is estimated that without greenhouse gases, the average global temperature would be about -18° C, which is too cold to support life as we know it. Mars has a thin atmosphere with a low concentration of water vapour and other greenhouse gases; temperatures on that planet are below -50° C. Venus, on the other hand, has a dense atmosphere with high proportions of radiatively active gases. The greenhouse effect on Venus maintains temperatures at above 400° C.

Many gases play a role in the natural greenhouse effect on earth, and the most important is water vapour. Water molecules are continuously cycling between the atmosphere and water surfaces, changing phase from vapour to liquid to solid. The evaporation, sublimation and condensation processes all involve substantial energy exchanges,

215

which drive much of the weather we experience. The average water content of the atmosphere changes little and is kept in balance through the evaporation-precipitation cycle. However, how much water the atmosphere can hold depends on temperature—a warmer atmosphere tends to hold more water vapour, providing a positive feedback mechanism to heat the atmosphere further.

Other important greenhouse gases are carbon dioxide (CO_2), methane (CH_4) and nitrous oxide (N_2O). These gases have a natural origin, but they are also produced through human activity. Carbon dioxide is released from combustion processes such as forest fires and through fossil fuel combustion in heating systems, engines and boilers.

The Greenhouse Effect

Fig. 13-3

A - sunlight travels though the atmosphere and strikes the earth's surface, which warms and radiates infared energy.

B - most of the infared heat is absorbed by the so-called greenhouse gases in the atmosphere. The gases warm and then radiate heat energy back to the earth.

C - burning forests and other fires release greenhouse gases, especially carbon dioxide, into the atmosphere. This is the natural greenhouse effect, which has given the earth an average surface temperature near 16° C.

D - additional carbon dioxide and other greenhouse gases are released to the atmosphere from fossil fuel combustion in factories and power plants and from transportation. This anthropogenic greenhouse effect is warming the earth and atmosphere at unprecedented rates.

Trees killed by the mountain pine beetle

Carbon dioxide is also produced in respiration and through the aerobic decay of vegetation. Methane is produced through anaerobic decay of vegetation and in the digestive tract of ruminants. It is the primary constituent of natural gas, is contained in coal beds and is a minor by-product of the combustion of carbon-based fuels. Nitrous oxide is primarily a product of combustion, though it is also released to the atmosphere when nitrogen-based fertilizers break down. All these gases have atmospheric lifespans that are much longer than that of water vapour, and this longevity is what makes them of particular concern as greenhouse gases because they tend to accumulate in the atmosphere.

There are other gases that have no natural form in nature but are only produced by humans. Substances such as CFCs, halons and several others are in this category and are potent greenhouse gases, in part because of their long atmospheric lifespans.

217

The Trend in Global Greenhouse Gases

Atmospheric CO_2 concentrations have been measured for less than a century. Measurements taken at Mauna Loa, Hawaii, since 1956 show a steady upward trend, with mean annual atmospheric concentrations increasing from about 300 parts per million (ppm) to the present value of 380 ppm. Similar trends are noted at Canadian sites and at other locations around the world. Analysis of the atmospheric gases trapped in glaciers and ice sheets provides a longer record. Ice cores from the south polar ice sheet have made it possible to reconstruct the atmosphere's CO_2 concentration extending back more than 600,000 years. Data from these cores shows that CO_2 concentrations were about 270 ppm prior to industrialization. During periods of major global glaciations, global CO_2 concentration levels were about 200 ppm. During the warm interglacial periods, including the present one, CO_2 levels attained peak values of 270 to 300 ppm. Methane and nitrous oxide concentrations show similar long-term trends. In summary, the present concentrations of all the key greenhouse gases are far higher than levels over the past 600,000 years.

Glaciers are retreating in a warming climate.

Atmospheric Carbon Dioxide Trends

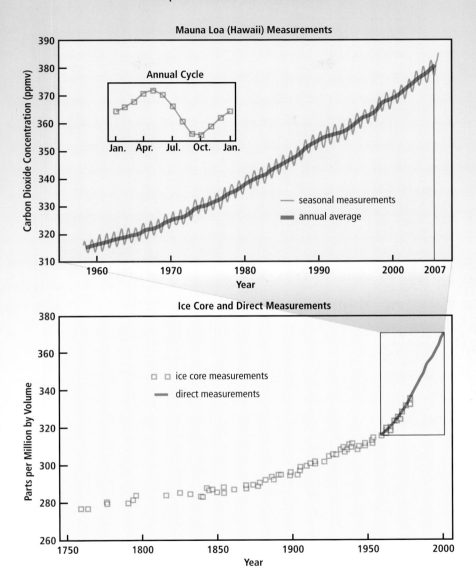

Fig. 13-4 Atmospheric carbon dioxide concentrations as determined from ice cores show a steady but modest increase from 1750 to 1950. Since 1960, instrument measurements and ice core data show the rate of increase has accelerated. Global values are now over 380 ppm.

The Changing Atmosphere

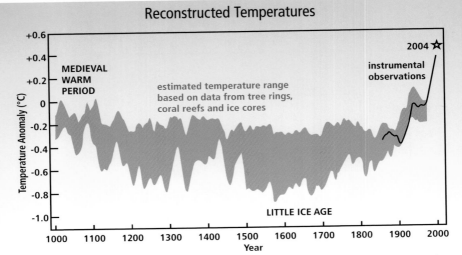

Fig. 13-5 The shaded region shows the range of uncertainty in temperatures inferred from tree rings, coral reefs and ice cores. The solid line is the instrumented record. All data reflect the global averages from 1961 to 1990.

Concerns about Warming

In 1896, while investigating the causes of periodic ice ages, Swedish chemist Svante Arrhenius quantified the relationship between the average concentrations of greenhouse gases and average global temperatures. He concluded that a doubling of atmospheric carbon dioxide concentrations would likely result in an average global warming of 5° to 6° C. Arrhenius realized that the CO_2 introduced by industrial emissions was in fact accumulating in the atmosphere, but he concluded that, given the emission rates at the turn of the 20th century, a doubling of CO_2 concentrations would be unlikely for 3000 years. In reality, emissions from new sources of industrial emissions are being produced at much greater rates than he envisioned, and a doubling might be expected to occur within 100 years.

There is clear evidence that the earth has undergone a steady warming over the past century. The scientific community has been working to confirm the climate trend and to determine the reasons. Careful examination of temperature data collected around the globe reinforces that a warming is underway. The temperature record shows that since the end of the 19th century, the average global temperature has risen about 0.6° C. This warming has not been steady. The instrument record shows that a period of warming from about 1910 to 1940 was followed by a short period of cooling until the mid-1970s. The warming since then has been steady. The variability of the trend has caused much controversy. Some brief cooling periods can be explained by volcanic eruptions such as Mount Pinatubo in 1991, but not all the variability has been explained. Nevertheless, the general

220

trend is decidedly upward, and most scientists agree that this warming is because of excessive greenhouse gas emissions.

Data has been compiled for the northern hemisphere, consisting of an amalgam of measured temperature records and temperatures inferred from tree ring growth, coral growth and ice core analyses. The period from about 1000 to 1900 showed a slight cooling. Over the past century, however, the warming has been remarkable (see figure 13-5).

Climate Changes in British Columbia

In BC, the temperature record extends back to the late 19th century. When annual temperatures are plotted, there is considerable variation from year to year, but the temperature data shows evidence of a warming trend. The temperature record for sites throughout the province shows elements of the same trends noticed elsewhere in the northern hemisphere. Figure 13-6 shows the global trend from about 1860 onward. In BC, there has been an overall upward trend in the annual temperature, with the northwest warming at a greater rate than anywhere else in the province. The coast and areas to the east of the Rockies have shown the least amount of change, with the rest of the province recording about 1° C of warming.

The precipitation data shows an upward trend in most areas of southern BC. During the period from 1929 to

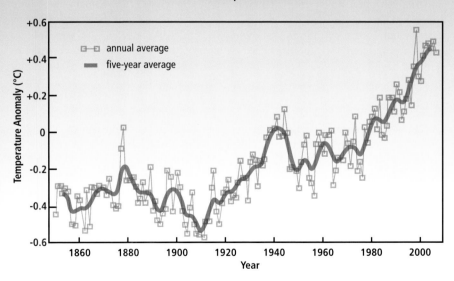

Fig. 13-6 The squares are average temperatures computed from thousands of surface stations around the globe. The data reflect averages for the period from 1961 to 1990. The smooth curve is the five-year running mean of the plotted data.

1998, the amount of precipitation rose at the rate of between 2 and 4 percent per decade. Not enough data is available to say whether this trend is also being experienced in northern BC. Climate models predict that precipitation will continue to rise slowly, within 10 percent, until 2050, and then increase by 10 to 20 percent after that. As temperatures warm, more winter precipitation will fall as rain. With rising freezing levels, winter snow levels will be higher in the mountains with less water from run-off available for summer use.

Future Climate

Predicting future climate is a challenging task. The climate system is composed of several components, including the atmosphere, the hydrosphere (oceans, lakes, wetlands and rivers), the earth's cryosphere (glaciers, floating ice) and the biosphere (including all living things). These components of the climate system interact and influence each other. Components of the climate system that have a physical science basis can be described by mathematical relationships. A representation of the climate system can then be made by means of a complex mathematical model formulated to operate on computers. This model is known as a climate simulation model. The equations depend on time, so the model can be used to examine how the climate system changes with

time. Scientists use this model to develop scenarios of the future climate.

Climate simulation models work much like the weather prediction models described in an earlier chapter, but there are several important differences. Weather prediction and climate simulation models both use the laws of fluid physics to describe the atmosphere and its changes with time. Climate models, however, take into account many additional processes that are not important for short-range weather forecasting. Weather prediction models start from an initial state of the atmosphere and run forward to a few days in the future. Climate models run forward for decades, even out to 100 years or more. The climate models run on a coarser scale than weather prediction models—usually with a horizontal grid spacing of hundreds rather than tens of kilometres—and therefore the climate detail is relatively coarse. All the complexities of the linkages between the atmosphere, the hydrosphere and the cryosphere must be calculated using relationships appropriate to the grid scale of the model. The number of calculations necessary to undertake a climate simulation can only be accomplished in a reasonable time frame by using the most powerful computers available. Only a few centres in the world have sufficient computing power and scientific expertise to undertake numerical climate simulations. Some of the more prominent centres include the Hadley Centre of the British Meteorological Office, the Geophysical Fluid Dynamics Laboratory (GFDL) of the National Oceanographic and Atmospheric Administration of the United States, the National Center for

> Love comforteth like sunshine after rain.
>
> —Shakespeare, *Venus and Adonis*

Atmospheric Research (NCAR) located in Colorado, and the Earth Simulator Center in Japan. In Canada, Environment Canada conducts climate simulation modelling with a group located at the Canadian Centre for Climate Modelling and Analysis in Victoria, BC.

The output of the climate simulation models can be displayed either in pattern form (i.e., maps) or by displaying the value of some climatic parameter that is computed by the model. Average surface temperatures over the globe, over a hemisphere or even over a region are often displayed. For meteorologists to have confidence that the models are correct, the models must accurately depict our present climate. A good test of their validity is to see how they perform in simulating past climate trends.

223

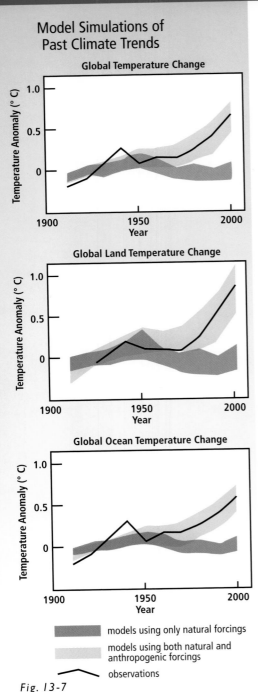

Model Simulations of Past Climate Trends

Global Temperature Change

Global Land Temperature Change

Global Ocean Temperature Change

models using only natural forcings

models using both natural and anthropogenic forcings

observations

Fig. 13-7

Figure 13-7 shows how 14 models have performed in simulations of worldwide surface temperature over the oceans and land services throughout the past century. Also shown is the average of observations calculated in the same areas over the same period. The ranges of the model simulations, shown as shaded bands, depend on the assumptions used in the formulation of the model physics. The graph shows that models best replicate the observed global temperature trend when increases in anthropogenic greenhouse gas emissions are incorporated in model simulations.

In 1988, the United Nations established the Intergovernmental Panel on Climate Change (IPCC) to investigate whether human activity is causing climate change. The emphasis over the past two decades has been on the role of greenhouse gases in changing the short-term climate. Climatic simulation models are used to undertake experiments using a range of greenhouse gas emission scenarios. Figure 13-8 shows the results of several emission scenario experiments undertaken using a number of climatic simulation models. The models show that there is a wide range of possible outcomes for the future climate, depending on the emissions scenario. The models project that by the end of the 21st century, the earth will have warmed between 1.1 and 6.4° C relative to the global temperature at the end of the 20th century.

British Columbia's Future Climate

The present trends in climatic observations together with climate model simulations suggest that the world's climate will undergo further warming during this century. The "Indicators of Climate Change for British Columbia" provincial government study, published in 2002, forecasts significant changes in the climate, sea level and ecosystems in the 21st century.

According to the document, the average annual temperature may increase by as much as 4° C, and precipitation should increase from 10 to 20 percent. With current projections of global ice melt and volume increase with warmer water temperatures, sea level may rise by up to 88 centimetres along parts of the BC coast. Many small glaciers in southern BC will likely disappear, with larger ones receding significantly. The potential loss of mountain snowpack may necessitate a greater reliance on artificial snowmaking to maintain BC's important recreational skiing and snowboarding industry.

Some interior rivers may dry up during summer and early autumn; this could lead to serious problems for communities that use rivers and stream as their water supply, and salmon migration patterns and success could change.

Multi-modal Averages and Assessed Ranges for Surface Warming

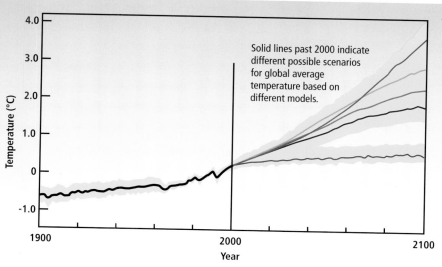

Fig. 13-8 The yellow shaded band shows the results of simulations of global average surface temperature from many different climate simulation models. The solid curve prior to 2000 is the observed temperature. Each solid line past 2000 is the average of model simulations for one emission scenario.

A thinning ozone layer poses a threat of sun damage to unprotected skin.

The mountain pine beetle and other forest pests will likely expand their ranges. British Columbia's boreal forests may be subjected to additional stress, and climate conditions favour the northward progression of grasslands. Changes in recreational resources could also be anticipated with a more arid climate. Wetlands might continue to decline, which would further stress migratory bird populations.

There will also be more extreme weather events, i.e., stronger winter storms on the coast and more violent summer thunderstorms in the Interior. Plants and animals will be severely affected with the changes. Extinctions are likely in some species, while others will have to adapt or migrate.

Because climate models typically use a fairly coarse grid, simulating climate

Impact of Climate Change

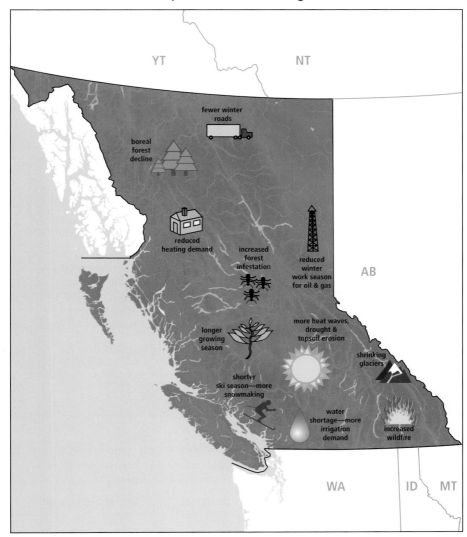

Fig. 13-9

changes at a regional scale is a major challenge. Regional climate models are being developed to deal with this issue and are being tested throughout western Canada. The outputs from the Canadian model are eagerly anticipated. More definitive studies will undoubtedly follow, providing better resolution of climate trends and impacts. This information will then allow for an improved understanding of the options available and the costs associated with accommodating a changing climate.

Glossary

accretion: the growth of a hydrometeor through collision with supercooled cloud drops.

aerology: the study of the atmosphere throughout its vertical extent.

air mass: an extensive body of air within which temperature and moisture at a constant height are essentially uniform.

albedo: the rate of reflectivity of incoming solar radiation at the earth's surface.

ambient air quality: air pollutant concentration levels for outdoor air.

anticyclone: an atmospheric pressure system that has relatively high pressure at its centre and, in the northern hemisphere, has winds blowing around the centre in a clockwise direction; also called, simply, a high.

azimuth angle: the angle in degrees from north measured in a clockwise direction.

biogeoclimatic: geographical area with a relatively uniform climate and associated vegetation.

blue jet: a weakly luminous blue-coloured discharge that moves upward from the top of a thunderstorm.

boreal: relating to cool northern climates and the associated plant and animal life.

cirrus: a principal cloud type in the form of thin or feather-like clouds in the upper atmosphere, made up of ice crystals.

Coriolis effect: the apparent deflection of moving objects as a result of the earth's rotation. Objects deflect to the right in the northern hemisphere and to the left in the south.

cumulus: a cloud in the form of detached domes or towers that appear dense and have well-defined edges; has a flat, nearly horizontal base with a bulging upper portion that resembles cauliflower.

cup anemometer: a wind-measuring instrument consisting of three or four hemispherical or conical-shaped cups mounted on horizontal arms that rotate around a vertical axis.

cyclone: an atmospheric pressure system that has relatively low pressure at its centre and, in the northern hemisphere, has winds blowing around the centre in a counterclockwise direction; also called, simply, a low.

discretization: the method of representing a continuously varying parameter with a discrete set of values.

downburst: localized area of damaging winds caused by air rapidly descending from a thunderstorm.

freezing level: the lowest altitude in the atmosphere where the air temperature is zero degrees Celsius.

front: the boundary region along the earth's surface that separates air masses; frontal types are characterized by the nature of the cold air mass, the direction of the movement of the front and its stage of development.

frontal lift: the upward displace-ment of an air mass of lower density by an air mass with a higher density.

frontal surface: the interface above the earth's surface that separates different air masses.

fulgurite: a glassy tube formed when a lightning stroke terminates in dry, sandy soil and fuses the sand.

geostrophic wind: the air motion that results from air pressure differences.

gradient: the change of a meteorological parameter per unit of distance.

graupel: a form of ice created in the atmosphere when supercooled water droplets coat an ice crystal.

Hawaiian high: high-pressure area that develops over the central Pacific Ocean about 1600 kilometres northeast of the Hawaiian Islands.

hydrometeor: a general term for any type of water or ice particle.

isobar: a line on a chart or diagram drawn through points that have the same barometric pressure.

jet stream: a zone of relatively strong winds concentrated within a narrow band in the atmosphere.

lapse rate: the rate of decrease of atmospheric temperature with height.

mares' tail: a cirrus cloud that has long, wispy strands extending from a tufted end.

mesocyclone: an updraft column of air with a diameter of 2 to 10 kilometres that rotates around a vertical axis in a large thunderstorm; the rotating column can lead to the formation of tornados.

mesopause: the atmospheric layer located at a height of around 90 kilometres, at the top of the mesosphere, where the coldest atmospheric temperatures occur.

mesosphere: the atmospheric layer above the stratosphere, characterized by temperatures that decrease from the base of the layer to the top.

N-region: the negatively charged region in the lower portion of a thunderstorm cloud

occluded front: formed when a cold front overtakes a warm front during the latter stages of frontal development.

Pineapple Express: a non-technical term for a strong and persistent flow of moisture and associated heavy rainfall originating in the waters near the Hawaiian Islands.

planetary (long) waves: a wave-shaped pattern in the westerly wind belt characterized by long length; four or five waves are typically found around the hemisphere, and the wave pattern generally moves slowly toward the east with a speed much slower than the speed of the westerly winds.

planetary boundary layer: the bottom layer of the troposphere in contact with the earth's surface.

plow winds: strong, straight line winds associated with the outflow of a strong thunderstorm.

p-region: the positively charged region near the bottom of a thunderstorm cloud.

P-region: the positively charged region near the top of a thunderstorm cloud.

pressure gradient force: the force acting on air caused by differences in air pressure; the strength of the force is proportional to the difference in air pressure.

psychrometer: an instrument that measures humidity using two thermometers, one of which has its sensing bulb covered by a water-saturated muslin jacket.

retrogression: westward movement of weather systems at middle latitudes, instead of the usual eastward movement.

red sprite: a large-scale luminous flash that appears directly above an active thunderstorm and extends from a cloud top to heights of about 90 kilometres; it is delayed a short time after a cloud-to-ground lightning stroke.

ridge: an elongated area of high atmospheric pressure on a weather chart.

rime: milky colored ice formed when super-cooled water drops freeze on contact with a surface and create a mass of tiny balls with air spaces between them.

shortwave trough: small-scale disturbances embedded in the upper air flow.

silver thaw: colloquial expression for a deposit of glaze on trees and other objects from a fall of freezing rain.

smog: originally defined as a mixture of smoke and natural fog, but now describes a mixture of air pollutants that are emitted by automobiles and industrial sources and are acted upon by the sun.

snow pillow: an instrument used to estimate snow pack by measuring the pressure exerted by a mass of overlying snow.

solar wind: a stream of charged particles flowing outward from the sun.

stepped leader: the column of highly ionized air that intermittently advances downward from a thunderstorm cloud and establishes a channel for a lightning stroke.

stratopause: the top of the stratosphere, characterized by a reversal of the rate of change of temperature with height.

stratosphere: the atmospheric layer from about 10 to 50 kilometers above the earth, characterized by temperatures that warm from the base to the top of the layer.

stratus: a cloud layer with a fairly uniform base.

streamer: an upward-advancing column of highly ionized air that moves from a point on the earth's surface toward a stepped leader.

suction vortex: a small-scale secondary vortex within a tornado core that moves around a central vertical axis.

supercell: a large, severe thunderstorm with rotating updraft that usually lasts several hours and produces heavy rain and hail, strong winds and occasionally spawns tornados.

synoptic scale storm: a large-scale atmospheric system or weather event that has a horizontal dimension of greater than 300 kilometres.

thermistor: a temperature sensor in which the electrical resistance varies in proportion to the temperature; the name is a combination of "thermal" and "resistor."

thermosphere: the upper atmosphere above 100 kilometers, characterized by low gas density and temperatures increasing with height.

tripole distribution: the characteristic distribution of positive and negative charge centers in thunderstorm clouds.

tropopause: the transition zone in the upper atmosphere between the troposphere and the stratosphere, characterized by an abrupt reversal of the thermal lapse rate.

troposphere: the atmospheric layer from the earth's surface to about 8 to 12 kilometers, characterized by decreasing temperature with height and substantial water vapor; the region in which most weather occurs.

trough: an elongated area of low atmospheric pressure on a weather chart.

Resources

Barry, R.G. and R.J. Chorley. (1971). *Atmosphere, Weather and Climate*. London: Butler & Tanner Ltd.

Bates, David V. and Robert B. Caton. (2002). *A Citizen's Guide to Air Pollution* (2nd ed.). Vancouver: David Suzuki Foundation.

British Columbia Ministry of Environment, Ecosystems Branch. "Ecology: Ecoregions of British Columbia."
www.env.gov.bc.ca/ecology/ecoregions/index.html

British Columbia Ministry of Water, Land and Air Protection. (2002) *Indicators of Climate Change for British Columbia, 2002*. (also available in print form)
www.env.gov.bc.ca/air/climate/indicat/pdf/indcc.pdf

Cessna Pilot Center. (1984). *Canadian Manual of Flight*. Denver: Jeppesen & Co.

David Suzuki Foundation. "Solving Global Warming. Impacts: British Columbia."
www.davidsuzuki.org/Climate_Change/Impacts/British_Columbia/

Etkin, David and Soren Erik Brun. *Canada's Hail Climatology: 1977–1993*. Toronto: Institute for Catastrophic Loss Reduction. (also available in print form)
www.iclr.org/pdf/Hail%20paper%20Revised%20.pdf

Drew, John. (1855). *Practical Meteorology*. London: John Van Voorst.

Fletcher, N.H. (1962). *The Physics of Rainclouds*. New York: Cambridge University Press.

Flohn, Hermann. (1969). *Climate and Weather*. New York: McGraw-Hill Book Company.

Fralic, Shelley, ed. (2003). *Wildfire: British Columbia Burns*. Vancouver: Greystone Books.

From the Ground Up. (1987). Ottawa: Aviation Publishers Co. Ltd.

Gdezelman, Stanley David. (1980). *The Science and Wonder of the Atmosphere*. New York: John Wiley & Sons.

Hare, F. Kenneth and Morley K. Thomas. (1974). *Climate Canada*. Toronto: John Wiley & Sons Canada Ltd.

Harrison, Louis P. (1942). *Meteorology*. New York: National Aeronautics Council, Inc.

Holton, James. R, Judith A. Cury and John A. Pyle (Eds.). (2002). *Encyclopedia of Atmospheric Sciences*. San Diego: Academic Press.

Intergovernmental Panel on Climate Change. (2007). "Summary for Policymakers." In *Climate Change 2007: The Physical Science Basis* (contribution of Working Group I to the Fourth Assessment Report of the Intergovernmental Panel on Climate Change). Cambridge: Cambridge University Press.

Johnson, Kent and John Mullock. (1996) *Aviation Weather Hazards of British Columbia and the Yukon.* Environment Canada.

Lange, Owen S. (1999). *The Wind Came All Ways: A Quest to Understand the Winds, Waves, and Weather in the Georgia Basin.* Environment Canada.

Stanley, George F.G. (1955). *John Henry Lefroy: In Search of the Magnetic North.* Toronto: MacMillan Company of Canada Ltd.

Meteorological Branch. (1964). *Weather Ways.* Ottawa: Department of Transport, Queens Printer and Controller of Stationery.

Natural Resources Canada. "The Atlas of Canada: Climate." atlas.nrcan.gc.ca/site/english/maps/environment/climate

Natural Resources Canada, Canadian Forest Service. "Canadian Wildland Fire Information System: Canadian Forest Fire Danger Rating System (CFFDRS)." cwfis.cfs.nrcan.gc.ca/en/background/bi_FDR_summary_e.php

Paruk, B.J. (1988). "Spring Blizzard 1988. Western Region Report." Unpublished document. Environment Canada.

Phillips, David. (1990). *The Climates of Canada.* Ottawa: Supply and Services Canada.

Rakov, Vladimir A. and Martin A. Uman. (2003). *Lightning: Physics and Effects.* Cambridge: Cambridge University Press.

seestanleypark.com. "Stanley Park Windstorm Damage, Vancouver BC, December 15, 2006."
www.seestanleypark.com/wind/wind151206a.htm

Shaefer, Vincent J. and John A. Day. (1981). *A Field Guide to the Atmosphere.* Boston: Houghton Mifflin Co.

Spiegel, Herbert J. and Arnold Gruber. (1983). *From Weather Vanes to Satellites.* New York: John Wiley & Sons.

Uman, Martin A. (1986). *All About Lightning.* New York: Dover Publications, Inc.

Weisberg, Joseph S. (1981). *Meteorology: the Earth and Its Weather.* Boston: Houghton Mifflin Co.

Williams, Jack. (1992). *The Weather Book.* New York: Vintage Books.

Williamson, S.J. (1973). *Fundamentals of Air Pollution.* Reading: Addison Wesley Publishing Co.

Zielinski, Gregory A. and Barry D. Keim. (2003). *New England Weather, New England Climate.* Lebanon: University Press of New England.

Wikipedia. "Hanukkah Eve Wind Storm of 2006." en.wikipedia.org/wiki/December_2006_Pacific_Northwest_storms

Index

Photo Credits

Ralph Adams / BC Government–Ministry of Environment 205, 206, 207, 211, 230-231; Cory Bialecki 185, 186, 187, 190, 191, 193, 195; Justin Brown 106a; Canadian Avalanche Centre, Revelstoke 134; Canadian Wind Energy Association (a Canadian wind farm) 83; Phil Chadwick 31, 32, 37, 40, 45, 53b, 62b, 92; Chris Cowan 84, 155, 167a; Digital Vision Ltd. 38, 218; Gabor Fricska 1, 5, 9, 29, 112; Harrison Hot Springs 135; Bill Hume 36, 48, 50, 51a, 161b, 178, 182; Christopher Hunter 152; iStockphoto.com / Hans Caluwaerts 212; iStockphoto.com / Slavoljub Pantelic 214; iStockphoto.com / Jason Verschoor 100; iStockphoto.com / Shane White 213; iStockphoto.com / Serdar Yagci 179; Jupiterimages Corporation 10, 34, 51b, 102, 181, 196, 197, 198, 201, 223, 226; jurvetson 97; Alister Ling 33a, 44, 47a, 72, 78a, 82, 108; Edward P. Lozowski 64; Luna04 78b; NASA 208; NASA / Veres Viktor 47b; NOAA Photo Library 60; NOAA Photo Library / Grant W. Goodge 41; NOAA Photo Library / Sean Linehan 160; PhotoDisc, Inc. 75; Public Domain 76; Public Domain (GNU Free Documentation License) 141b; Public Domain (U.S. Government) 139; Mike Roberts 109b; Heather Rombough 35a; Ed Tyson 144; Mark Wilson 105; Randy Kennedy 180; Rob Pigott 4, 8, 25, 39, 43, 46, 49, 52, 54, 65, 90, 95, 96, 99, 118, 138, 145, 149, 166, 188, 199, 217; Shawn Pigott 33b, 184a; Steve Coe 66; Terri Lang / Kelowna National Services Office, Environment Canada 35b; United States Air Force / Senior Airman Joshua Strang 80-81; Vlad Bilkun 200; www.victoriaweather.ca (UVic School-Based Weather Station Network) 30

© Her Majesty the Queen in Right of Canada / Pacific Storm Prediction Centre (Vancouver), Environment Canada, 2009. Reproduced with the permission of the Minister of Public Works and Government Services Canada: 173, 183, 184b

© Her Majesty the Queen in Right of Canada, Environment Canada, 2009. Reproduced with the permission of the Minister of Public Works and Government Services Canada: 141a, 153, 156, 157a, 157b, 158a, 158b, 159, 161a, 162, 163a, 163b, 164, 165, 167b, 177

© Her Majesty The Queen in Right of Canada, Environment Canada, 2009. Reproduced with the permission of the Minister of Public Works and Government Services Canada / NOAA: 170b, 171

Authors

Rob Pigott started his career in Atmospheric Sciences in 1968 with Environment Canada in Ottawa. For the next 24 years he served as a weather observer and presentation technician in various locations throughout BC. He then spent 13 years at Kelowna Mountain Weather Centre as a weather forecaster specializing in fire weather and mountain forecasting. Rob retired in 2005 and then started Enviro–BC Weather Services. His company is currently under contract to provide fire weather support to the BC Ministry of Forests and Range. In the winter, he keeps busy preparing mountain forecasts for several interior BC ski resorts. He also writes weather columns for local newspapers and teaches fire weather forecasting courses.

Rob and his wife Laura live in Courtenay and have a combined total of five grown children, one beautiful granddaughter and an entertaining cat named Cici who meows like a bird.

Bill Hume started his career in atmospheric sciences in 1971 with Environment Canada in Toronto. He attained a Master's degree in meteorology from the University of Alberta in 1975 and has spent most of his working career in Edmonton and Calgary. He has experience in most aspects of meteorology including weather forecasting, climatology, air quality science and atmospheric research. He has taught meteorology at the University of Alberta and served on several national committees including the Canadian Meteorological and Oceanographic Society. Bill and his wife Judy have two married children and live in Edmonton.